VOICES FROM THE APE HOUSE

VOICES
from the
APE HOUSE

Beth Armstrong

Trillium, an imprint of
The Ohio State University Press
Columbus

Library of Congress Cataloging-in-Publication Data
Names: Armstrong, Elizabeth (Elizabeth Lynn), author.
Title: Voices from the ape house / Beth Armstrong.
Description: Columbus : Trillium, an imprint of The Ohio State University Press, [2020] |
    Summary: "A memoir from an influential Columbus Zoo gorilla keeper and
    conservationist"—Provided by publisher.
Identifiers: LCCN 2019034834 | ISBN 9780814255711 (paperback) |
    ISBN 9780814277652 (ebook)
Subjects: LCSH: Gorilla—Ohio—Anecdotes. | Zoo keepers—Ohio—Anecdotes.
Classification: LCC QL795.G7 A76 2020 | DDC 599.88409771—dc23
LC record available at https://lccn.loc.gov/2019034834

Cover design by Amanda Weiss
Type set in Adobe Garamond Pro
Cover photo by Tom and Jan Parkes
♾ The paper used in this publication meets the minimum requirements of the American
National Standard for Information Sciences—Permanence of Paper for Printed Library
Materials. ANSI Z39.48-1992.

For Bongo and his brethren

And to Mom-mom, who taught me that every family has a story

worth the telling and the knowing

I laughed the first time I was asked "Why a gorilla?" To me the choice was all but self- evident. What creature could be more impressive and authoritative spokesperson for the nonhuman community of life?

—DANIEL QUINN, AUTHOR OF *ISHMAEL*

If human beings are defined by a sense of self-awareness, then without a doubt that is something that great apes share with us. When you see a gorilla, it is a gorilla being. And you realize that humans are not the only self-conscious and aware creatures on the planet.

—IAN REDMOND, INTERVIEW WITH *THE INDEPENDENT* (REPUBLISHED IN *GORILLA GAZETTE*)

They [gorillas] are brave and loyal. They help each other. They rival elephants as parents and whales for gentleness. They play and have humor and they harm nothing. They are what we should be. I don't know if we'll ever get there.

—THE LATE PAT DERBY, CO-FOUNDER OF THE PERFORMING ANIMAL WELFARE SOCIETY

CONTENTS

ILLUSTRATIONS

These are my recollections.

All of the stories in this book are true—as best as my memory serves—other than the prologue. Each of these stories I either witnessed myself or they were told to me by my fellow keepers and are noted as such. But the order of some events may have been changed for the narrative to flow more smoothly.

The prologue is based solely on my imagination, on what I envisaged a young gorilla might have experienced when captured from the wild. The brutality, cruelty, and trauma of capture are true enough. What those infants would have witnessed, would have felt, and experienced surely are true enough as well.

A final note on the text:
Some when reading this book will accuse me of being
anthropomorphic—I plead guilty.

PROLOGUE
1958

The morning began as all his mornings do, peering at the world from the safety of his mother's enfolding arms, listening to the soft steady beat of her heart, smelling her comforting musky odor. The birds begin their cacophony and Colobus monkeys cough and bark at one another. In the forest below, a forest elephant walks by unseen, and duikers on their delicate hooves mince their way through the foliage.

In nearby nests, other troop members begin to stir; his father greets his sleepy troop with a low rumbling "naa-hummm," signaling that the day has begun. His mother's body reverberates against the infant's as she rumbles back. Other group members call their greetings and then begin their slow descent from their night nests, infants clinging to their mothers' bellies, juveniles meekly following their mother, aware that their days of nest sharing with her are soon coming to an end. The silverback reaches the base of his night-nest tree and begins to walk in a westerly direction toward a familiar patch of wild ginger, where they will spend the morning feeding. While the adults forage, the infants and juveniles play chase and chest-beat in mock imitation of their elders, sometimes losing their balance in their comical attempts to look fierce.

After leisurely feeding for several hours, the troop settles in for their midday naps, which offer an opportunity for the youngsters to use the huge silverback as king of the hill, chasing one another up, over, and around him. Sometimes he casually grabs the nearest infant and tickles her until she gulps for air, desperately trying to get away. Inevitably all the youngsters will pile on the male laughing until collapsing in a heap of contented sighs and groans. When the silverback has had enough, he softly cough-grunts at one of the nearest youngsters until they all settle down. The infant boy, enclosed in his mother's arms, watches all of this . . . learning.

His mother is the dominant female and has successfully raised several other infants. His younger sister sits nearby patiently waiting for an opportunity to hold her brother when he wanders away from their mother. She tenderly plays auntie to him, practicing the skills she will need in the future to be a good mother. Eventually their mother gently retrieves the infant from his sister and settles in for her nap. The group lay scattered about in quickly constructed day nests, making contented belch vocalizations, passing gas, or reaching out a hand in an attempt to subdue a still unruly infant into submission.

The afternoon rain starts gently falling, making plopping sounds as it falls on the understory leaves, the gorillas lulled by the sound. But something shifts, a subtle change; the forest becomes quiet, the birds suddenly silent. The silverback sits up listening, then stands stiff-legged, lips pulled taut in warning before emitting a piercing alarm bark, which is abruptly cut short. He stands swaying slightly, a perplexed look on his face quickly replaced by one of sadness before he slumps forward. The infant watches . . .

Then his mother flees with him in her arms, followed by his youngest sister screaming in panic, leaving a trail of fear diarrhea behind them. His eleven-year-old brother closely brings up the rear, issuing short sharp alarm barks as they desperately try to find a way out of the encircling men. Other troop members can be heard screaming, only to suddenly break off.

The infant's mother's heart is racing; she emits a strong, acrid odor, a fear smell. She becomes quiet, maybe recognizing that if they are to survive she must be invisible. Her eldest son acting as a rear-guard falls even further behind, the mother turns, glimpsing her son as he stands upright, mouth open, screaming at their attackers before being cut down by a single shot. She runs on, the infant jostled by her frantic movements, his sister is whimpering, desperate to stay close to their mother.

Suddenly, the female stops. Human faces, white and black surround them. She looks directly at a white face as he raises a long gun to his shoulder and points toward her. She feels the searing pain only for a moment and then she is on her back, her eyes open looking through the forest canopy to the storm clouds passing over above. She knows now she can do nothing to protect her children. The infant scrunches down flat against his mother's body, his face buried deep in her neck in a futile attempt to become invisible, his eyes closed tightly in denial of what is happening. His mother is still breathing when she feels the infant boy being pulled from her body. Her son is screaming, his lips pulled back, cough-grunting, trying to bite the hands grabbing him, but she is helpless to comfort him. She feels as if her heart is breaking. In his fear, the infant is shitting himself and anyone in close proximity. He is smacked for his impudence. His hands and feet are tied to a long pole then hoisted off the ground, carried by a porter at either end, his wrists and ankles screaming in agony as the weight of his body feels every step and misstep. He does not know where his sister is.

When they stop later, the infant is transferred to a hastily constructed wooden box where he curls up. While the infant tries to sleep, his sister peers through the underbrush past the men sleeping around the fire to her brother. She works her way around the clearing until she is closer to him. She can hear him whimpering in despair and vocalizes softly to him both in hello and goodbye and disappears into the forest. He raises his head briefly and peers into the forest.

Her story will be different from her brother's. Eventually she will find another dispersed member of her troop, an adult female with an infant who was able to escape. They will wander for days through the forest until they come upon another troop, led by a silverback they have never seen before. Both females will integrate into this troop but not without paying an enormous price. The adult female will watch the new silverback grab her infant giving it a swift death-bite. By killing the infant the male will ensure that the female will begin her reproductive cycle again. He will then impregnate her and produce his own offspring, which he will then protect. Gorillas know the rules: there is nothing to be gained genetically for the silverback to raise another's infant.

His young sister, having always been high-ranking in her natal troop due to her mother's status, will now be the lowest on the totem pole. She will have to find a way to navigate into the good graces of all those within this new and unfamiliar troop. Her younger brother will have to find a way to navigate a completely different world from hers.

The infant curls up in a ball, his arms covering his head, and peers out into the night. Having always had his mother's body to cling to for warmth, he shivers from cold and fear. He is hungry and thirsty, having not eaten since early that morning. Nearby, some of the men awaken, making noises at one another, what sounds at times like laughter. He watches until one gets up and approaches the box. The infant curls more tightly into a protective ball, the man pokes him with a stick, and the infant curls tighter into himself as if to disappear, not making a sound. The man pokes harder this time, and the infant screams, which seems to be what the man wants. He turns laughing to the others. In response, another man gets up and approaches quickly, grabbing the stick, breaking it, then throwing it away, all the while making noises at the stick-man, which sound to the infant like the cough-grunting his father used to make when angry at a troop member.

The young gorilla is carried from the forest on his long journey to the capital of Cameroon, Victoria, a loud bustling city. The smell of diesel, wood smoke, and that peculiar odor that emanates from humans—combined with the noise of the city—overwhelms him. He learns to be cautious. When food and drink are placed in his makeshift enclosure, he waits until the humans leave before he gathers himself to warily approach it. The man who had been kind to him realizes that the open sides of the crate trau-matize the infant so he has burlap sacks placed over the crate, allowing the infant to retreat into his own world. Whether he does this from true kindness or because the infant is an investment, nothing more than a com-modity, who knows? But when he approaches the infant, he speaks gently and makes clear to the others that they are not to tease or be cruel to the in-fant in any way. Then the youngster is placed in a larger wooden crate and placed below deck on a cargo ship to begin his long voyage to the United States, to his new and alien life as a captive gorilla.

I feel more comfortable with gorillas than people.
I can anticipate what a gorilla's going to do,
and they're purely motivated.

—DIAN FOSSEY

I

APE HOUSE

It is always the smell I remember. Opening the back kitchen door to the Ape House lets loose a wave of gorilla—a thick deep pungent odor that envelops me as I unlock the door to let myself in. The aroma rolls out, dissipating into the early morning cool. By end of day, I will smell like a gorilla, their perfume clinging to my clothes, my hair, my skin.

The gorillas hear me. The click of the huge padlock alerts them that I have arrived and another day has begun. I hear their welcoming vocalizations, a deep rumbling, each greeting as distinct as the individual animal that elicits it. I cross the kitchen to the solid wooden sliding door that opens to the back keeper aisle and unlock it. As I slide it to the right, it reveals Colo, the first gorilla born in captivity, the matriarch of our gorillas, sitting in her usual dignified way at her barred back door. She has half a grapefruit husk perched snug on her head, perfectly balanced on her sagittal crest. A crown, if you will. She looks at me in such a dead-on serious way—I turn away to hide my smile. She looks silly, so sincere and thoroughly imperial. How does she pull it off?

"Good Morning, Colo," I say as she carefully watches me. I have never been one of her favorites. Our initial meeting on my first day in the Ape House didn't go well and I don't think she has ever forgotten or forgiven me. She tolerates me, but I don't think she's overly fond of me.

Colo is one of the most beautiful gorillas I will ever meet. The contours of her face are heart shaped; her brown eyes are penetrating with an unusual rim of light gray around each iris. Her hands and feet are long and narrow. Her fingers and toes are equally long. In a word, they are elegant. Her coal-black skin is flawlessly uniform, her thick hair coat healthy and a deep black. She has grayed out as she has gotten older, but she has none of the reddish hair on her head that many of our other females have. She reminds me of a New York City matron from a well-to-do family: always smartly turned out, hair and make-up in place, wearing the perfect Chanel suit.

I turn back to the kitchen, start a pot of coffee, soak the eight food bins filled with vegetables and fruit, and then grab the vitamin bottles and begin my walk down the back aisle, distributing vitamins to each of the gorillas.

This daily morning walk ensures that the gorillas receive their vitamin C and multivitamin tablets. But more importantly, it allows us to check on them, to give them a quick look-see to make sure all is OK. Over the years, this walk will reveal much heartache and joy. It will make known that in the depths of the night their lives went on without us, that babies were born, beloved troop members died—a daily reminder that we really are periphery to their complicated lives. We may be charged with "keeping" them, but they never will be "ours."

We keepers are the fortunate few in this world to bear witness to their lives, to every once in a while be asked to join in a bout of play or chase, to be accepted as an observer watching the careful tenderness of a silverback touching his newborn son for the first time, or to unexpectedly come upon a boisterous round of tickling between two usually reserved adults, and backing away quietly so as not to interrupt them. We few hear the sad mourning call after the death of a troop member, the haunting refrain wafting through the Ape House, its tender tendrils looking into every nook and cranny as if searching for the lost group member. We are, quite simply, privileged.

The back keeper aisle is long and narrow, bright with the glare of lights casting a yellowish hue when turned on but quite dark and cave-like when off. The wall to the right is made of cement block. All the trappings of gorilla keeping are there: hoses, shovels, pitchforks, scrub brushes, and squeegees hang neatly in their assigned spaces. On the left are the five barred doors, each an entryway into a fourteen-by-twenty-six-foot enclosure.

In the summer of 1982, head keeper Dianna Frisch is giving me my first tour of the building, quietly explaining who is who, where the cleaning supplies are, what my duties will be. The residents are Baron Macombo, or Mac as he is known, and his great-granddaughter Cora, who is in an adjacent enclosure but due to the design they are unable to see one another. The middle houses five black-and-white Colobus monkeys. Twenty-six-year-old male Bongo and his mate, Colo, share the last two enclosures. Dianna and I stop by Colo's back door. It is my first meeting, and even to my novice un-trained eye Colo is stunningly striking.

Colo leans against the barred door as she softly vocalizes to Dianna. Dianna talks to her in a quiet soothing voice. It is evident even to me, with no gorilla experience, that there is a great and enduring friend-ship between them. Colo then looks past Dianna, noticing me for the first time, and cough-grunts directly at me, holding my eyes with her own. I am thrilled. A gorilla is actually "talking" to me. So, I do a right back-atcha cough-grunt, thinking to myself, "My, isn't she being friendly!" In all fairness to my naiveté, I was raising a trio of baby raccoons for a local wildlife rehabilitation center at the time and vocalizing to them in their trilling purr fashion is an important part of their rearing process.

But now I am in a gorilla's world, where I soon learned that a cough-grunt is one of the most basic vocalizations in their repertoire, sounding like a hard "oo, oo, oo" rhyming with "new." It is the most commonly used method of showing obvious displeasure at another, be it gorilla or human. It is a way of saying, "I don't know you, don't like you, don't want you near me." But I dig myself even deeper. Along with my incongruous cheerful cough-grunt, I stare directly into Colo's eyes, considered to be another direct threat in a gorilla's social lexicon. Later back in the kitchen after our tour, Dianna kindly and gently let me know about my gaffes, but for now I am blissfully unaware.

This two-punch combo forever marked me in Colo's estimation. Looking back, I honestly believe Colo was startled, truly taken aback at my well-intentioned but mistaken response. She did a not-so-subtle double take, with a look of indignant disbelief on her face.

Me, I'm still completely oblivious to the fact that I have probably set the tone for our future relationship. I amble on to the next enclosure, and there he is, Bongo. He is livid at my intrusion in his world; his body tense and rigid, arms stiff, and his irate face taut with lips pulled in tight. He is majestic in his anger, and I am at once intimidated and in absolute wonder of this animal standing before me.

Here, right in front of me, is the animal I observed as a child when we made our annual visit to the zoo in summer. I was disturbed even then by his intentional studied stare, blankly looking into the distance past the visitors as if he could somehow will himself out of these small confining enclosures, as if he could spirit himself away from the deafening voices of the hundreds who came daily to rudely stare and then shuffle on to the next exhibit. I remember sitting quietly on the cement bench along the back wall of the public aisle observing him. I always wanted to stay longer in the Ape House, especially when the crowds left, because it seemed to me, if only for a moment, that his body relaxed, his demeanor became less defensive, and his shield came down. I felt, even as a youngster, a sense of unease at the injustice of this animal's life.

But now, no more than a few inches from him and only a barred door separating us, I am not prepared for how massive he is, how angry his eyes are, and just how beautiful he is all at once. Then I feel something else—as if I have come home.

2

CHILDHOOD

What is it that shapes us? I have given much thought to this lately. The clay of childhood that is pieced together, the memorable experiences that are added, fashioned, and spun into the beings we become. From my earliest childhood, there were people and events that began the process of forming me, pointing me toward becoming who I am today.

Did our next-door neighbor, the soon to be fully fledged veterinarian Dr. Jim Hugenberg know that his kind gesture of taking a four-year-old with him on his weekend rounds at The Ohio State University would lay a foundational slab of clay that would become my life?

At some level, the experience was excruciating for me as I was so painfully shy in the company of anyone other than my family members. I remember that part a bit, but I also remember the excitement of holding newborn puppies mewling and squirming in my lap, their silky smooth coats and fat bellies, their puppy breath smell. I also recall walking by Dr. Hugenberg's side in the large animal barn with its distinct sweet scent of fresh hay and the whiff of manure, hearing a cow's moo, the friendly nickering of horses. Of leaving his side and running ahead and around a corner only to be confronted by a massive bull with a ring through his nose, pensively eyeballing me.

I am also convinced we are shaped by places, that our childhood landscapes inform our sensibilities. My parents bought their first home when I was five in a neighborhood with an eclectic mix

of charming bungalows, Colonials like our house, or small Cape
Cods like my grandparents' home just four blocks away, all built in
the 1920s, '30s, and '40s. Our house was situated smack dab in the
middle of the two primary north/south roads in Columbus, Ohio,
but far enough from each that it offered a quiet respite from the busy-
ness of a city and its consequent noise. In the distance and especially
at night when sound always seemed to carry, I lay in bed listening to
the sorrowful call of the train whistle as it neared crossings at the eastern boundary of our neighborhood.

A series of green islands divided the road at the end of our block,
creating a one-way traffic pattern in either direction. It was a peaceful
street, our green zone, and all the neighborhood kids used it to walk
or bike to the pool in summer, to the nearby recreation center in fall
and winter, or to and from our Catholic school. It was a safe and protective womb with little or no car traffic.

Sugar and red maples lined our streets, offering up their sticky
helicopter-like seedpods that we broke open and wore on our noses
in spring and carefully blew on them when dried out to create a buzz-
ing whistle sound. In summer, the trees created a rainforest canopy
over our road when in full leafy bloom. Our house had a maple in between the sidewalk and street curb, and its trunk was our home base
for nightly games of hide-and-go-seek in summer. It was the favorite
tree for climbing with a perfect low-hanging branch that we used to
hitch ourselves up into its protective cover. Oak trees were scattered
throughout the neighborhood, towering monoliths high above the
maples. In fall, the oaks offered up their leaves for building forts and
their acorns for ammo when we played capture the fort from dawn
to dusk. Interspersed throughout the backyards were peeling, sloppy-
barked sycamores that suddenly transformed when stripped of their
leaves in winter, their smooth bone-white trunks rising up distinct
against the brilliant blue skies of a crisp, cold winter day. The beauty
of the seasons surrounded and informed all of us, as we are a generation that found our happiness outside, in the company of the smells
and sounds of nature.

It is the height of spring; the loamy soil is still damp and the morning chilly as I walk to school. Robins make their way back and forth,

scurrying along in their busy comical gait plucking up bloated reddish-pink worms for their broods back at their nests. They are on a frenzied mission of caretaking.

By afternoon it has warmed up considerably, my navy blue cardigan securely tied around my waist over my tartan school uniform as I walk home. I am in the fourth grade, and we have the most astonishing teacher, a lay teacher by the name of Miss Penny Ury. She is young, a bit on the plumpish side, pretty with light reddish hair, and she has a passion for teaching.

She has taught us about gauchos in Argentina, what it's like to surf in Hawaii, how Eskimo houses are built. Miss Ury brings in food from every geographic region we have studied thus far, opening our minds to different cultures and regions of the world through the pleasure of taste. She planted the seed that learning is not linear but holistic. She showed us that while cultures may reside in a place, culture is more than geography. It is also smells, languages, local customs, and art. During this magical year of learning, she taught us to memorize every state of the union by singing, in our enthusiastically sweet eight-year-old voices, "Fifty-Nifty United States," which we endlessly sing on the playground or on our daily walk home.

Our teacher is ingenious in her approach to gathering our attention, holding it, making us think independently, exploring the world on our own. She has given us the greatest gift of all—curiosity. Without preaching, Miss Ury teaches us to be open to others—to embrace and luxuriate in our differences.

But perhaps the greatest adventure is the chicken egg project. Miss Ury has brought in an assortment of jars—each displaying the developmental stages of chicken embryos. Floating in formalin, they line a shelf along the wall for our young minds to explore, and rather than being put off by these strange, otherworldly creatures, we are fascinated.

We clearly see the growth and changes from blob-like creature to the lengthening of the body, the distinct formation of eyes, and the growing head. And if that is not enough, she has brought in an incubator. For the next three weeks the students will be in charge of turning the eggs three to five times daily, checking the temperature, charting everything on our tablet. Miss Ury has created a team of enthusiastic scientists.

It is during this year of wonder, on this glorious spring day on my way home from school and looking forward to my weekend, when something catches my eye, a break in the pattern of leaves and dark soil. At the base of a tree I see broken blue eggshells. Nearby I see movement, a creature from another time, something that looks pre-historic, something that looks very much like our floating embryos. Featherless and naked, encased in a pink translucence, I can see its heart beating rapidly through its parchment thin skin. It is ungainly with a huge abdomen, the bulbous eyes not yet open, the wings nothing more that prickly prongs. Its cavernous yellow beak opens when it senses movement nearby, the scrawny neck seemingly unable to support its large head as it searches to and fro looking for sustenance. I am enthralled. I pick it up and it is cold to the touch so, wrapping my hands around it for warmth, I carry it home like some sort of precious cargo.

With my mother's help, we create a comfortable nest in a small cardboard box, line it with cotton scraps, and add a hot water bottle wrapped in a towel for warmth. Mom goes to the drugstore to buy an eyedropper to feed it. These were the days long before there were wildlife rescue centers, so we are on our own, but my father knows a veterinarian, so he calls her, seeking advice on what to feed and how often.

The bird seems to rally after a couple feeds, and we chart our feeding schedule, every half hour throughout the day. It starts its peeping whenever it senses movement in my bedroom. I am thrilled while feeding this creature but equally delighted when it seems sated after a feed and settles down for a rest.

It seems to thrive for the first day or so, but on the third day, I wake to ominous silence and peering into the box, I see that its tiny body is still, its mouth slightly ajar. I examine it for a cause but can see nothing. Wrapping it in a clean dish towel, we bury it in the backyard. I am sad, but something has happened, my passion for living things, for animals, has been sparked.

Growing up in the 1960s was complicated. Our childhoods were touched by magical things like lightning bugs, games of kick the can, and endless languid summer days but always with the backdrop of the Vietnam War. We lived in a decade of music so diverse that the likes have yet to be seen again. Constantly streaming from our transistor radios were the Temptations, the Supremes, the Beatles, the raspy voice of Dusty Springfield, the smooth tone of Glen Campbell, and the sensual sound of Brazil's bossa nova. This soundtrack was juxtaposed against nightly casualty reports from Walter Cronkite, of antiwar demonstrations, riots, and the assassinations that shattered our hopes, leaving us confused and bereft, in a state of melancholic longing for what could have been.

My three brothers and I excitedly look forward to watching Walt Disney's *Wonderful World of Color* every Sunday night. We get our baths done early without our mother's cajoling. We are scrubbed clean, smelling of soap, our wet hair lying damp on our pajama collars. We gather in front of the TV in eager anticipation, our sibling quarrels set aside for once, and wait for the opening music and the appearance of Tinker Bell. We sit cross-legged on the floor, riveted by tales of wild animals, of a family that raises a sea lion in their California home, of *Old Yeller* or *The Yearling*, stories that will break our hearts and bring us to tears.

And one magical Sunday night, Elsa the lioness made her appearance in our living room. To this day, I remember very specific scenes from the movie *Born Free*: Elsa's mother being shot, her cubs being found, and Elsa softly "pffting" while rubbing up against her beloved Joy Adamson. I felt the heartache and Elsa's loneliness when her littermates were shipped off to a European zoo. I remember Elsa terrorizing the African staff with her playful but frightening ambushes.

The movie stayed with me. I rode my silver and teal bike down our street, no hands, relishing the warm summer evenings as I sang the song "Born Free" at the top of my lungs. I was captivated with Joy and George Adamson's life out in the bush, with their rock hyrax, Pati-Pati, who lived at their camp, and with their relationship to the wild. At the library, I checked out the book of the same name and spent endless hours poring over the black-and-white photos, reading and rereading. Another spark had been added to the slow burn.

We had a cat when I was a child, aptly if not unimaginably named Kitty. Kitty was not a particularly affectionate cat. Later in life, she softened around the edges, but early on she was a tough street cat, given to spending most of her time outdoors. So it was with much surprise when something extraordinary happened between us.

Kitty was pregnant, judging from her expanding belly and her nipples now pink and evident. One morning she came into the house and ran straight up to my bedroom where I was reading on my bed. She was restless, her meows insistent, and I realized she was going into labor. So I cleared the bottom drawer of my dresser, lining it with newspaper and old rags. She waited patiently on the floor beside me. When I finished, she climbed in, examined her surroundings from corner to corner before she settled in, kneading the bedding while purring loudly.

I stayed with her for a bit to make sure she was comfortable, then decided to give her some privacy, so I headed downstairs. After only a few minutes, Kitty came to the top of the landing and began to meow insistently. At first I thought it was my imagination, but then I realized she wanted me back up in the bedroom with her. My face flushed with pleasure and pride as I headed back up the stairs to my room as she scampered on ahead, once again settling in.

I watched her give birth, talking softly to her as her body convulsed with contractions between each kitten's entrance into the world. I marveled at the gray tabby, the yellow tabby, and the unexpected solid black kittens as they drew their first breaths and mewed for their mama. I reveled in the smell of them and her as she brought forth and cared for each one, patiently cleaning away the afterbirth. I heard the distinct suckling sounds as each kitten found its way to their mother's teat. I watched in wonder at the perfect curl of the kitten's pink tongues when they fell off the nipple in a milk-induced stupor and continued to suck on air as they drifted off to sleep.

3

CLARITY

Seemingly random events can play a significant role in defining a direction in one's life, events that gently and sometimes more obviously catch your attention, prompting a visceral response, a moment when casual interest turns to passion. There is finally an instant when all that nettlesome longing transforms into clarity, and you know within your heart, at your very core, what it is you must do with your life.

I moved back to Columbus in 1979 at the age of twenty-two and stayed with my father and stepmother as I tried to find my bearings. One evening, I was sitting in their family room watching the news. I had started to leave after the broadcast when a local TV magazine program, *PM Columbus,* came on with a segment highlighting a baby gorilla named Cora. As if drawn by some invisible string, I immediately sat back down. Cora was being raised in the Columbus Zoo nursery, and I was absolutely captivated. At that very moment, I not only wanted but also felt the need to work with gorillas.

Fueled by this glimpse of Cora, I applied to be a volunteer at the Columbus Zoo. Perhaps apply is too formal a word—I showed up. Back then, every volunteer started off in the Children's Zoo barn. It was while hosing the petting area, feeding animals, and helping to monitor the interaction of the visitors with the barn animals that I found a sense of peace and orderliness, and a sense of achievement after a day of hard physical work.

During that same summer, an infant gorilla, Roscoe, was born. He was being cared for at Children's Hospital on the east side of the city. Roscoe had contracted salmonella, a bacterial infection, while being raised in the Children's Zoo nursery with another infant gorilla named Kahn. Kahn was born three days before Roscoe, but became ill, dying several days later as a result of the same infection. I don't recall if I was asked to work with the infant Roscoe or if I approached the assistant director and volunteered—probably the latter. I was juggling several minimum-wage jobs and had very little money to speak of. I drove a beat-up gas guzzler of a car. So I asked the assistant director if I could be paid for that one day a week of working with Roscoe, in order to cover my gas costs as well as hospital parking fees. And voila, simple as that, my first paid job as a keeper!

Swathed in a deep sense of pride, I walked from the hospital parking lot into the building and found my way down through the basement labyrinth to where Roscoe was. I felt a sense of purpose, a sense that a piece that had been missing in my life had just slipped into place with a firm click. Roscoe was being closely monitored due to his critical condition and was in an incubator. I garbed up in a medical gown, surgical mask, and gloves. I monitored his respirations, fed him, and kept track of his urinations and stools, noting their consistency. It was the early days, and we were not allowed to remove him from the incubator, but I somehow knew in my gut that what he needed was a chest to lie on, to hear another's heartbeat, to feel the comfort of someone's rhythmic breathing lulling him to sleep. But I didn't presume to do so and instead placed my gloved hand through the opening of the incubator onto his back for what I hoped would be some measure of comfort. I felt his steady breathing softly rising and falling under my hand as I breathed in his unique baby gorilla fragrance beneath the talcum powder odor of my latex gloves and beyond the antiseptic smell of the hospital. I was transfixed and thrilled beyond all measure.

After having worked in the barn and with Roscoe, then after being laid off after the summer season, I applied to The Ohio State University. I was still scrambling to pay my bills and was working in a bar/restaurant. Of my several roommates at the time, my favorite was Donna, a full-time student finishing up her last year at Ohio State. Donna worked for a local Catholic parish cooking dinner for their three priests five nights a week, a job she would cede over to me when she graduated. She was a steadying influence on me, had a very matter-of-fact approach to life, and a smile that absolutely illuminated her face.

Donna's fiancé, Jeff, lived out of town, but on one of the rare weekends he visited Donna, he was kind enough to walk me through the Pell Grant application. With papers spread out all over the living room floor, Jeff patiently assisted me in filling out all the forms and attaching the requisite tax returns. The result was being awarded a grant for the 1980–81 academic year. The grant would pay for my tuition for three quarters at Ohio State.

It was a disaster. I found the campus huge and intimidating, the number of students overwhelming. Just finding a parking place seemed insurmountable. Most harmful of all, I found myself envying those students who drove new cars, had their housing paid for, and their tuition and books covered. I wished for myself another life where I could attend a small college campus, where I had a family that could cushion me from the trials of paying bills, and of constantly dealing with broken-down cars that shamed me to drive.

Envy is one of the seven deadly sins for a reason: it is corrosive. If I could talk to that young woman in her early twenties today, I would say this: "Don't, just don't. You are good enough. You can do this. Have faith, you will find a way, never doubt that you can be anything you set your mind to." But looking back now, I realize that I have always had a thin fragile layer between myself and the coarseness of the outside world—it's that shy kid deep within me. It is as if my nerve endings are sensitive to every changing nuance. Being that sort of person, the large, kinetic Ohio State campus rattled me to my core.

In the winter of 1982, feeling deflated and with my confidence still at an all-time low because of my poor academic performance the year before, I heard about an upcoming lecture at the zoo by Marsha King. Marsha and I had begun volunteering at the Children's Zoo barn around the same time, and now she was raising an infant chimpanzee for the zoo. I attended her lecture, reigniting my passion of working with animals. The doors reopened and I walked straight through them and began volunteering once more at the zoo.

I still worked at the bar/restaurant Crazy Horse Saloon and after many years of waiting tables, I had grown weary of the bar business. What had once been a fun (and reckless) alcohol-fueled life had definitely lost its sheen. I started to work in the kitchen as a cook with head-cook Michael Chillik. Mike was a consummate and discerning observer of the human experience, hilariously funny when commenting on humankind's quirks, always on point, but never unkind. I also changed from the night shift to the day shift when possible, thus avoiding the nightly drinking crowd. It meant less money but more peace of mind. Michael always worked the day shift as well. Every once in a while he might stay for one beer after his shift, but that was pretty unusual.

One early May morning, the bartender called me from the kitchen telling me I had a phone call. I was laughing at some smart-ass remark Mike had just made as I pushed through the swinging kitchen doors and grabbed the phone. "Hello, this is Beth." It was Don Winstel, the assistant zoo curator, asking if I would be interested in a job at the zoo, a forty-hour a week seasonal position in the Children's' Zoo. The job would start immediately, ending in mid-September.

That call, that opportunity changed my life. It was like being thrown a lifeline while treading water in a hopelessly vast ocean. I got off the phone stunned. Something had shifted, as if the sun had come out in one glorious moment after a particularly gray, dreary day. I walked back into the kitchen, in search of Michael who was in the freezer unloading a delivery and I told him the good news. We immediately headed to the bar and ordered a shot of tequila, a testament to Michael's support for me. And with that, I was on my way.

4

PRIMATE BEHAVIOR 101

It was in the Children's Zoo where I unwittingly began my apprenticeship as a primate keeper. The Children's Zoo encompasses the barn in the south end with its goats, sheep, chickens, and a beautiful dappled-gray horse named Farwan. The north end has all the exotics: red panda, kangaroo and wallaby, prairie dogs, porcupines, and a variety of primates that included squirrel monkeys, Golden lion tamarins, cotton-top marmosets, a pair of white-handed gibbons, and a troop of Japanese macaques, commonly known as snow monkeys. Between the north and south ends is a low white stone building that houses our office break room, the kitchen and cooler for food prep, and the infant nursery with its large plate glass windows for public viewing.

Working with snow monkeys was my introduction to a truly social group of primates, and they absolutely fascinated me. Snow monkeys are the most northern living species of primates with pinkish red faces and a thick gray-brown hair coat. Medium-bodied, females weigh around twenty pounds and the male twenty-five pounds or more, and both have impressive canines. Most people will recognize snow monkeys from *National Geographic* photos as they sit in Japan's thermal hot springs, a monkey hot tub, surrounded by snow—their dense fur surrounding their face, crystallized with ice, with an expression of utter bliss and relaxation on their faces.

But entering the Snow Monkey exhibit was a frightening rite of passage. A large A-frame enclosure with a low stone perimeter wall about three feet high with the remaining structure enclosed in chain-link fencing, the exhibit housed a group of five adults, two males and three females, and their offspring of varying ages. The substrate was sand so a keeper had to go in and clean with a rake and shovel, picking up feces and discarded food. After cleaning, the monkeys are given fresh water and their food spread throughout the exhibit.

As a novice keeper, I was asked to go in each Sunday morning, the start of my workweek. This meant that I went into the exhibit alone, negotiating the highly complex social world they inhabit by myself. The adult male, Matsu, was dominant, with Beta being the subordinate and backup male. The top-ranking female was named Beta's Bitch (for a reason), then came Long Tits (you get the picture), and last was Pretty Face (self-explanatory).

As I would leave the keeper building with two large red food buckets in the morning, it was Long Tits who served as the group's lookout, sitting high in the enclosure and letting loose with a vocalization that alerted the other group members that breakfast was on its way. Once inside the small inner safety enclosure, I set the food buckets down and gathered cleaning utensils before entering the actual monkey's exhibit. Matsu was the key to my safety, although I didn't know it initially. He was a strong but gentle leader, reassuring group members that all was well within their world. He broke up fights and, in general, was a calming presence.

As part of his daily routine, Matsu would climb down the chain link directly across from me and begin his walk on the stone wall before dropping down on to the sandy substrate. He would walk behind me while giving a reassuring vocalization to the group members (and me) before climbing the vertical fencing again. He was on patrol and would be until I exited the enclosure.

I was absolutely thrilled to work the north end of the Children's Zoo. I didn't want to appear to be a troublemaker or a whiner, so I was reluctant to question the existing protocol—sending a completely untested keeper with no proper primate knowledge in to a potentially dangerous situation. I was alone in the exhibit, if things got out of hand, it could go south in a hurry. Inexperienced though I was, intuitively I knew that this was nuts. After several nervous and sleepless Saturday nights, I finally approached the head keeper, and we came up with an alternate plan.

On Sunday mornings, Don Wright, head of the Children's Zoo barn, would accompany me into the exhibit just to watch my back. I will forever be grateful to him. Don is unflappable, and in time I came to realize that if you were in a tough situation, it was Don you wanted by your side. He instilled confidence simply by his calm steadiness. Eventually, having a pair of keepers in the snow monkey enclosure became standard practice, much to my relief.

In time, the substrate of the enclosure was changed to cement, which meant hosing was possible, and having a hose is a nice backup if there is any aggression, subtle or obvious, toward the keepers. A directional spray of the hose may create just enough diversion to allow time to exit the exhibit if trouble arose.

Once cleaning and feeding were done, the other keeper and I would exit to the small auxiliary safety enclosure, upturn the now empty food buckets, and take a seat for fifteen minutes or so to watch the monkeys' behaviors. Feeding time in any primate world is a quick glimpse into their social hierarchy. Matsu and Beta's Bitch, the dominant male and female, and their offspring always got first dibs on food. Being a subordinate member affects your offspring not only in a physical way—less access to food in general and choice food items in particular, which means your nutritional needs are not being met—but being of low rank has psychological effects as well. For instance, because Pretty Face was the lowest-ranking female, the dominant female routinely picked on her. Even if Pretty Face was not being targeted at any given moment, she was always on edge. If Beta's Bitch was having a bad day, Pretty Face would pay the price. As a result, Pretty Face routinely overgroomed her offspring. Most had bald heads or patchy hair coats. If uneasy, she would seek out her infant, holding him tight to her chest as she nervously plucked his hair. But Long Tits, the middle-ranking female, was left alone for the most part. She raised her kids, kept her own council, and didn't get into anyone else's business. She was also a more mature female, which seemed to weigh in her favor.

Dominant male Matsu was an assured and temperate leader, intervening in squabbles only if necessary. But Beta's Bitch could be trouble, as if she needed to prove a point both to lower-ranking group members and to us, the keepers. Snow monkeys, when aggressive, will do a quick lunge of the head and shoulders toward another group member or a keeper, with a cough-like vocalization, while bobbing their head to catch your eye. It's comparable to walking down the street when a person looking for a fight takes an aggressive stance right in your path. Direct stares are routinely used as a threat. Snow monkeys have a way of stretching their forehead around their eyes, tightening their eyelids to create an exaggerated stare. A direct stare is very focused and has intent behind it (you see the common thread here). All of these are meant to engage another group member in a potential altercation or to force an individual to display submissiveness. The number one rule in the snow monkey exhibit was never ever let them catch your eye, most especially, Beta's Bitch.

Sometimes Beta's Bitch was in an aggressive mood, just because, attempting to make eye contact even before we entered the enclosure. If that happened we had two options: either go back to the Children's Zoo office to grab a pair of sunglasses, which would mask our eyes while allowing us to keep a surreptitious eye on her, or, on extremely bad days, wear sunglasses and also summon Don Wright for additional backup.

One day while cleaning with fellow keeper Dusty Lombardi, an incident happened that scared the bejesus out of both of us. Dusty was hosing while I was about eight-feet away picking up detritus. Directly overhead was the horizontal support beam, and it is quite high. One of the infants was crawling across it when she lost her grip and fell landing directly in between us, making a distinct and alarming thump on the concrete. The sound alone caused us to groan inwardly. Since this infant happens to be top-ranking Beta's Bitch's youngest offspring, and as Beta's Bitch never needs an excuse to go after the keepers, when the infant hit the dirt (so to speak), Dusty and I both reacted with our own "Oh, we are so fucked" under our breath. And, because the infant had landed between us, you guessed it: the monkeys would automatically think we were responsible for hurting one of their own.

The infant appeared to be OK, but because it clearly has the wind knocked out of it, we know there will be a momentary pause before it starts to scream, which in turn will bring Beta's Bitch and the other troop members to its rescue. We quickly back out, moving in a slightly comical, silent-movie jerky kind of way. We made it out just before the frightened and panicked infant began screaming bloody murder. Needless to say, Beta's Bitch was not happy with us, as we collapsed in nervous laughter once outside their enclosure. She picked up her youngster, did a perfunctory once-over check, and then headed over to us now safely ensconced outside and gave us a vociferous piece of her mind.

Six weeks after being hired in the Children's Zoo as a seasonal summer worker, a permanent part-time keeper position opened up in the Ape House: thirty hours a week with all the responsibilities of a full-time position and without the benefits. No insurance, no paid holidays or full-time pay, but to me it sounded like heaven.

While I had been volunteering two years before in the Children's Zoo, a keeper position had opened up in the Ape House and I distinctly remember being filled with such a sense of "how lucky" for the gal who got the job. It was akin to a physical ache—my wanting to work with gorillas. It was an indicator of where my heart lay. This same girl was leaving so the position opened up and I applied for the job. But again perhaps "applied" sounds a bit too organized; back then it was pretty loosey-goosey around the zoo. I simply stopped by the Ape House to let them know I was interested.

Head keeper of gorillas, Dianna Frisch, came down several days later to the Children's Zoo and picked me up in her golf cart. We drove around while she asked me questions. I don't remember anything specific; I think Dianna was just trying to get a feel if I would be a good fit with her and Bill Cupps, head keeper of cheetah and bears, which was the other half of the department. Lucky for me, I apparently passed the test. I began almost immediately.

5

BEGINNINGS

In addition to working with gorillas and chimps, my other duties included taking care of cheetahs, otters, peccaries, hyenas, and an assortment of bears, including polar bears, covering for head keeper Bill Cupps on his days off. Bill imparted sage advice when training me on polar bears: "Beth, never forget you are nothing but lunchmeat to them," a warning I never for a moment forgot. Bill was an extraordinary keeper and relied on a common-sense approach to working with the animals under his care. It was Bill who pioneered the cheetah-breeding program at Columbus—using only behavioral methods—unlike other zoos that were routinely injecting their females with hormones. Bill watched, read about, and learned how cheetahs behaved in the wild and then simulated those very same things in captivity. He let cheetahs be cheetahs.

Several years later, Bill built a cubbing den for our female polar bear and convinced management that in order for her to successfully raise her cub, she would have to be placed in the den prior to giving birth and left in there for a long period of time after the birth, thereby mirroring what wild polar bears do, only emerging from their ice dens when their cubs are several months old. And Bill succeeded in 1988 with a healthy mother-reared cub. It would be another twenty-eight years before another polar bear raised her young at the zoo. I felt so fortunate to work with these different species and to learn so much from Bill, but truthfully it is the three days each week that I worked exclusively with gorillas that I looked forward to the most.

Sunday was Dianna's day off, and it was my first day of working the gorilla building by myself. Bill had left to take care of the animals in his area, and I spent an inordinate amount of time sequestered in the kitchen consuming even more cups of coffee and chain-smoking in order to build up my courage. When I could avoid it no longer, I took a deep, tremulous breath and grabbed the plastic food bin.

I am a complete unknown to the gorillas, a stranger, and I had less than a favorable beginning a week ago during my introductory walk down the back aisle where I met Colo. Am I friend or foe or simply something to be tolerated? And I am touching their food, one of the few things that bring pleasure and diversion into their otherwise sterile environment. Another strike against me is that one of their previous part-time keepers used to juggle their favorite food item, oranges, in front of them purely for the entertainment of the visiting public. Suffice it to say, he didn't last long in the department. Here's something all gorilla keepers will tell you, if you really want to piss a gorilla off—other than the most obvious, you know, like cough-grunting at them while looking them directly in the eyes—be cavalier with their food.

I had been given some tips and warnings from both Dianna and Bill about what to expect when feeding: "Bongo has a tendency to throws things, look out for the uncooked sweet potatoes." "Colo will attempt to sucker you in, she'll try to get you to come closer with an extended hand. It's a set-up; she can reach all the way out to her shoulders in a split-second and grab you. Oh, and by the way she's a spitter." "Just remember that when you try and separate Bongo and Colo for feeding, Bongo can throw the door in between their two enclosures open with a flick of his index finger and skewer you with the steel rod bar."

Armed with those words of encouragement, I make my way down the back aisle with the rectangular plastic food bin propped on my hip like an infant child. The base of all the enclosures begins with a short cinderblock stone wall with black vertical steel bars embedded into it. The vertical bars are approximately four or five inches apart. About half way up, a horizontal bar adds additional support and runs

the length of each front. The aqua blue painted floors are worn and
chipped; smooth light teal glazed cinderblock walls make up the remaining three sides, and each enclosure is topped with steel bars. A
cement water bowl is in one corner and a concrete bed raised several
feet off the floor is in the opposite corner. This is what makes up each
individual "room." They are completely and utterly sterile, with no
type of bedding to soften them. In some cases, there is no way of even
offering hay, as there are no options available for shifting animals out
of their enclosure to remove soiled bedding. It is a cold, unforgiving
concrete shell they live in.

All goes well my first day. Bongo does throw things at me. I am
not fooled by Colo's attempt to draw me closer. I emerge back in the
kitchen unscathed, even while ducking any manner of projectiles. I
am aware that I am the closest thing for them to vent their frustration
and anger at, and a calm realization descends on me that I am just
fine with that.

Gorilla keepers like nothing better than telling stories. Every keeper
has a story that moved them more than others, a special moment that
solidified their commitment to and love of working with gorillas, an
instant that floored them with its significance and continues to inform them even years later.

In September 2016 my husband and I were in a hotel pub on the
Isle of Skye in Scotland, after a long day of exploring its most ancient
and starkly beautiful landscape. Outside, the rain continued as the
misty evening descended on the Munros before the sun briefly illuminated them to an even deeper gold before the gloom returned. It
was good to be sitting with a drink in hand in front of a fireplace,
when in walked an eclectic group of fellow travelers, led by a black-
robed nun, their host. One member of the group, rail thin in her late
sixties and with a short, severe gray haircut, wore a fetching teal-and-
black kilt and starched white blouse. She sat down next to us and
introduced herself and then her fellow travelers as they slowly made
their way over from the bar and joined us in front of the fire.

We began chatting about our respective lives when my husband
casually mentioned that I used to work with captive gorillas. And so
the stories began, unheralded. My favorite stories of Bongo slip easily

from my tongue, as these strangers sit beguiled, not by me, but by the gorillas themselves.

It is my sixth week in the Ape House. I have learned that gorillas react to subtle and not so subtle facial expressions and body language. Paying attention to and studying other group members is just a part of the complex social pact within any given gorilla troop. It's a social survival mechanism, allowing them to maneuver through their hierarchy and find their place. It is no different with us, the keepers. The gorillas watch us, study our movements, sometimes wait for vulnerabilities, and at times solicit interactions. They are discerning when it comes to inexperience, and they can spot a bullshitter miles away.

In order to hide my nervousness, I learn to develop a studied nonchalant poker face while feeding, always keeping an eye on them with subtle glances. I recognize Bongo's method of quietly gathering a food arsenal at his feet, which could be used as missiles if he is of a mood. I perfect my food tossing technique between the vertical bars while standing several feet back out of Colo's reach, rarely missing. More importantly, I learn not to look stricken when I do miss and need to move closer in order to retrieve the fallen food item to then give it another try.

On this particular day, Bongo comes over to his usual spot at the front of the enclosure, sitting sideways to me, resting his back against the solid wall, quietly waiting for his food. Both his face and body are relaxed. I toss his orange in. He vocalizes a rumbling belch vocalization, a drawn out "mmm-wahhhh," sounding very much like a person clearing his throat. This vocalization can mean any number of things, including "Here I am. How's it going?" or "Here I am. Where are you?" or "I'm content. Life's good at the moment." It's a delicious sound, full of quiet goodwill.

He picks up the orange and gently pushes it out. Confused and slightly suspicious, I think to myself, "Hmmm is he trying to draw me in closer?" But I study his face, and it is relaxed, unperturbed. I pick up the orange and cautiously toss it back in. He immediately "mmm-wahhhhs" again, picks it up, and pushes it back through where it plunks on the keeper aisle floor directly in front of his enclosure. Guardedly, I watch him as I pick it up once again and lightly toss it back in.

FIGURE 5.1. Bongo

He now seems to be getting a little frustrated; his expression brings to mind an exasperated eye-roll you might see in a human when slightly annoyed while trying to explain something rather simple to another who doesn't get it. Bongo quietly gathers himself and ever so patiently pushes the orange back out—again. After this third attempt, it finally dawns on me; I realize Bongo is asking me to peel and section the orange, his favorite food item. I do as he wants, and he watches me closely, supervising, making soft pleasing vocalizations as if to encourage me on. I carefully lay each perfect half-moon segment along the horizontal bar. Bongo then leans forward, slurping up each individual orange section. Afterward, he gives a satisfied contented "mmmm-waahhh."

Even though I am a novice, I know in my heart, without a doubt, what he has just given me. He has given me his most treasured gift, his trust and acceptance. I walk back to the kitchen elated and humbled at the same time.

6

CHIMPANZEES

I have not yet mentioned the two chimpanzees that we also house in the Ape House. A male named Coco and his companion, Emily, live in the two twelve-by-thirteen enclosures adjacent to the kitchen. Coco is very handsome with a beautifully proportioned face and a wiry compact body. His arm and leg muscles are pronounced and fully developed. Emily is a sweet, gentle soul with saggy sad eyes in a long hangdog face, her enormous ears protruding, and Coco simply adores her.

Chimpanzees are fast as lightning and can cross an enclosure in a matter of seconds, arms and hands fully extended for a grab. Much has been said about their strength. Most sources say they have four to five times the strength of humans. So unlike gorillas, which are more reserved, chimps can be loud, reactive, and demonstrative, especially the males. But they can also be introspective and quite polite.

There is a story that circulates throughout the keeper ranks that pretty much captures the different personalities of the great ape species. If a hammer were accidently left in an enclosure, a gorilla would keep his distance, spooked by something out of the ordinary. A chimpanzee would use the hammer for aggressive displays and beat the crap out of the enclosure. An orangutan would use the hammer to slowly and methodically dismantle the enclosure, thus making her escape.

The door we used to separate Coco and Emily for their morning feeding and cleaning is the same iron bar system we have at Bongo and Colo's enclosures. It is a collapsible steel rod with a pin that holds the door in place when shut. When I try to close the door, Coco can easily grab it from within, flinging the door back open, essentially pitching me back and forth if I were to unwisely hold on to the steel rod too tightly.

Because of the configuration of the enclosures with only the back barred doors affording the keepers visual viewing (there are no strategically placed peepholes), we are essentially "blind" when trying to shut these types of doors. So the best bet is to quickly eyeball where the chimps are, hoping and praying they stay put while we attempt to slam the door in place. Because it is so precarious and dangerous, I try to shut the door when they are unaware and least expecting it. I act nonchalant as I go about my morning business, as if engrossed in something else, when I seize the moment to shut the door. This of course pisses Coco off because it is so unexpected that it startles him. Imagine being in your own home and something falls off a shelf crashing to the floor, or when a thunderclap occurs on a sunny day. I'm sure it is the same feeling for him and Emily. First it gives them an unexpected fright, followed immediately by an angry, knee-jerk reaction.

I think it is an exceedingly bad idea to house gorillas in the same building as chimps or bonobos for that matter. Both species are much more vocal than gorillas. But chimpanzees in particular can be aggressive when aroused, banging and displaying when agitated or insecure. They are loud arbiters of their social lives.

As soon as I arrive in the morning, Coco begins his excited pant-hoots, culminating in a banging of his cement walls with both hands and feet and rattling his back doors. He repeats this several times before stopping. When I finally have their food ready and open the door to the keeper aisle in front of their enclosures, he begins again in full vocal mode, literally bouncing off the walls, starting from the elevated sleeping platform with one leap across the enclosure to where I am, then full foot-stomping both walls, then back up to his sleeping platform, only to begin all over again. But for his second round of displays, he ever so smoothly picks up a pile of soft chimp poo, which he then deftly flings at me before finally settling down for his morning fruits and vegetables.

Emily, on the other hand, sits quietly in her opposite enclosure waiting for her meal. Sometimes she will take her stalks of celery and carrots and rather than eating them, she places them end to end, creating a perfect circle around herself—a facsimile of a nest. Both chimps and gorillas in the wild routinely create day and night nests to sleep in. Think of how we feel with a cozy blanket when taking a nap—creating a nest is analogous to that, giving one a feeling of comfort and warmth. It breaks my heart that giving nesting materials is not a part of our daily husbandry practice. In some cases, this is due to logistical hurdles, but in other cases, we actually could have given something. But at the same time, there is real concern, especially in these back holding areas where we have no ability to transfer them to an outdoor area in order to remove soiled bedding.

After the morning feeding, I typically hang my head in the sink and wash my hair and face. I learn early on to keep my mouth tightly closed when feeding the chimps. I sponge off remnants of chimp poop from my uniform before getting on with the cleaning. Recently, my former roommate and I were talking about my early days at the zoo; she looked at me in her usual unswerving manner, shaking her head while stating the obvious, "Beth, you do know you reeked when you came home—every single day."

Sometimes after feeding, Emily comes over to me, indicating she wants to groom my arm. Any type of freckle, mole, or imperfection is irresistible to her. She solicits me with pursed lips, a soft face, and soft vocalizations with a begging motion, asking for my arm. I give it to her willingly and she ever so gently pulls it through the bars—this is a dance of trust. She carefully holds my arm, mouthing the freckles while frequently looking up at me as if to ascertain my permission, all the while making happy lip-smacking sounds.

It would be ages before I finally found a way to work with Coco in a way that worked for both of us and that gave him some measure of calm. Eventually I discover that if I begin immediately vocalizing with Coco as he began his vocal displays it somehow soothes him. It is as if we are forming some sort of harmony together, as if my pant-hoots are somehow taking the air out of his own vocalizations. Or perhaps, it is that we are communing together through a shared song. If I calmly and matter-of-factly match his vocalizations, he slowly calms down, quietly comes over to the front of his enclosure, and waits patiently for his food, while making the most delightful feeding vocalizations.

7

REBELS

In 1587 the Scottish parliament felt compelled to pass an act "for the quieting and keeping in obedience the inhabitants of the Borders, Highlands and Isles." This act was partially directed at those pesky border clans living on the ever-fluid margins of southern Scotland and northern England. I bring this up because a special mention was made of the Armstrongs: "On the border were the Armstrongs, brave men, somewhat unruly, and ill to tame." Since discovering this statement and feeling an inordinate sense of pride in my troublesome ancestors, I started thinking again about what shapes us.

Through our accumulated experiences of childhood and adolescence, we learn and eventually are cast into the beings we become, just like young gorillas or other primate species. But what of genetics or what of some weird sense of collective memory? What of something unseen, something intangible like our ancestor's circumstances and histories? Can elements of their lives somehow make their way down through the ages to eventually manifest in me, a great-great-great-granddaughter?

In addition to the Armstrongs on my father's side of the family, we are descended from the Wynnes of Pennsylvania on my mother's paternal side. My eleventh great-grandfather was Dr. Thomas Wynne, personal physician to William Penn. Grandpa Wynne was a Welshman (ah, yes, the Welsh, another thorn in the side of the English monarchy throughout the ages). He was a devoted Quaker and as such was jailed six years by the English for his religious beliefs. In 1682 Wynne accompanied William Penn on the ship *Welcome* on its voyage to America. King Charles had given an enormous amount of land to Penn Jr. in reparation for money borrowed from Penn's father. Thomas Wynne, would serve as the speaker for the first two assemblies of Pennsylvania as well as a go-between negotiating with the Lenape Indians.

A renowned Quaker himself, William Penn Jr. envisioned a land free of religious persecution and the willingness to tolerate and work with others of differing beliefs. A pretty radical and, one might say, rebellious approach.

Quakers felt that they were "called by God to work for justice and peace." They did not believe in rituals and representatives of God, such as priests, but rather believed that God could be found in the everyday, and was inside each of us. Some might go so far as to say they were seventeenth-century hippies. So on one side we have the marauding Armstrongs, one of the infamous reiver clans of southern Scotland, and on the other, Welsh Quakers who sought peace in all things, all rebels in their own right.

We get our physical looks from our parents and from our ancestors. In old photos, we might see a common curve of a smile, the shape of a nose or the color of eyes. I clearly see commonalities in a family photo of my great-grandparents' fiftieth wedding anniversary party in 1953. In the background, surrounded by her seven sisters, is my beloved grandmother. She is smiling into the camera, and it is my smile sure enough, the same grin I see in my twenty-year-old niece. But we also seem to get specific mannerisms. I see that in my husband's granddaughter, in her sometimes-loping walk so similar to his. I think we also pick up certain ways of being in the world, certain perspectives. Could it be possible that my forefathers and foremothers, that troublesome Armstrong clan and the quite certitude of the Quakers, have somehow influenced me?

Bucking the system is not easy and truthfully does little for one's career. It makes people nervous, this voicing of opinions. But for someone like me who often felt outraged at what I thought were injustices in the zoo world, how fortunate was I to work at a zoo that not only allowed but encouraged those of us in the Ape House the autonomy to shake things up.

When training new gorilla keepers, I say to each and every one, hoping to instill in them this fundamental thought, "You are their voice. These animals can't waltz over to the zoo director's office or attend a national zoo conference to articulate their concerns. It is an intrinsic duty for every single keeper to push back against the status quo and to push for improvements. Your most important job is to be their advocate."

Recently I saw a posting on a zoo's Facebook page, highlighting the fact that its gorilla keepers wear surgical masks everyday—all day— with the explanation that gorillas are susceptible to all transmittable viruses we may unintentionally expose them to, such as colds or the flu. I get and accept that premise. And that is not to say that if you have a cold or are not feeling well, that you shouldn't wear a mask (you should) and limit your contact with gorillas. More effective, simply don't come into work that day.

I first saw the use of masks as a part of a gorilla husbandry program twenty-plus years ago at another zoo and found it alarming then. What I find most disturbing about this latest trend that now seems to be embraced by most, if not all, zoos in North America, is that we the keepers and staff now hold all the cards. We ask the gorilla's every day to do things that they may very well not want to do. They are vulnerable to us, to our whims. And now, with masks, they can't see our faces as they try to decipher our intent or motivations.

As much as we care about them, worry about them, and work hard to make their lives both individually and collectively better in captivity, we are in essence in control, so therefore the balance is already skewed. As Violet Sunde, a respected colleague and fellow gorilla keeper from Woodland Park Zoo in Seattle, once wrote, "Zoo gorillas are keenly aware of their captive condition. Gorillas share almost 98% of their DNA with humans and are highly intelligent, individual emotional and psychosocially complex beings. Even in the best zoo environments, captivity imposes unnatural stresses that would not be endured in the wild state. Major among these pressures is the fact that their lives are controlled day in and day out by humans."

All keepers will tell you that the most fundamental tool in a keeper's repertoire is the ability to read a gorilla's body language, and most especially their facial expressions. Each gorilla has a distinctive face, a unique body type, and a specific way of comporting himself that is unique to him alone. I think what I find so bewildering, and frankly troubling, is that by covering our faces we are not giving them the full respect they deserve. There is something intrinsically insulting about this approach. We want their cooperation, their trust, their part in this lopsided partnership, yet we don't allow them access to our faces, the window into our own thoughts and moods.

When I saw this posting extolling the virtues of wearing masks, I informally canvassed several of my fellow gorilla keepers. They all agreed that wearing masks on a daily basis was unnecessary. They most especially cited the unfairness of hiding our faces. Here are a few of their comments:

> "I've always been opposed. I felt that masks obscured too much of our faces, making it hard for the gorillas to read non-verbal cues."
>
> "I consider masks inconvenient and unnecessary for every-day tasks. Gloves, yes."
>
> "I agree with everyone. The zoo's protocol used to be 'got a cold flu etc., work another section till you are better.'"

As I was preparing to write this book, I reread Jeff Lyttle's *Gorillas in Our Midst.* Jeff's book brought home to me the role of rebels in creating the Columbus Zoo gorilla program, especially the integral case of Ohio State veterinary student Warren Thomas. He was employed as a keeper in the late 1950s. Thomas took it upon himself to ignore the clear and unequivocal mandate of then zoo director Earl Davis to not let gorillas Millie and Mac in together, for fear of injuries. Instead, Thomas routinely opened the doors and allowed them access to one another overnight and separated them when he came back in the morning. This was a pretty ballsy move considering little was known about gorilla behavior in 1955–56. But because Thomas chose to go with his gut, Colo was the result. She was born on December 22, 1956, the first gorilla born in captivity. All else flows from that one rebellious act.

In 1978 a young Jack Hanna was hired by Metro Parks director Mel Dodge to oversee the Columbus Zoo. As with many zoos of the day, there was room for myriad improvements. Jack's personality was one of possibilities, not rebellious for rebellion's sake, but rather a positive and always inclusive "Why not?" outlook. He appeared unruffled by his fellow zoo directors, many of whom shook their heads, bemoaning and criticizing his every move. Although in private conversations with him I did get the impression sometimes that he did worry and that perhaps he might have pushed it a bit too far in some instances. He tells these stories like a relieved kid, his eyes wide open with hands spread out, like someone who had just gotten away with something in his neighborhood, not anything malicious or unkind, just something that skirted the conventions of the day.

The wonderful thing about Jack was that he never seemed motivated by career advancement. Making decisions weren't weighed down or filtered through the screen of how it would affect his career or what the latest zoo trend was. Instead, he asked, "What can we do here?" His attitude allowed for an incredibly creative environment in which to work. He backed our ideas and nothing was too crazy or considered off the table. But if you came up with an idea, it was expected that you would see it through from start to finish—fair enough. This approach was not exclusive to the gorilla department; it was a zoo-wide phenomenon, evident in Bill's cheetah breeding program, in the bird department's many accomplishments, and in the nursery's unique approaches to raising infants.

And don't take Jack's affable manner for naiveté or mistake his southern accent and sometimes "golly" refrains, as some sort of southern stereotype—don't underestimate him. When Jack made a decision, he was like steel, willing to take the heat for it. I saw this firsthand in Guatemala in 1990 while we were investigating the possibility of supporting a nongovernmental organization (NGO) that was rehabilitating confiscated wildlife. This would be our first official foray into supporting a field conservation project that involved a large matching grant. We were in a restaurant in Guatemala City, having a surprisingly adversarial meeting with a woman from another unrelated NGO who wanted our funds as well. She was badmouthing the group we were considering supporting and continued to do so incessantly. We all tried to remain polite and were at a loss as to how to proceed without being overtly rude, but she was riding roughshod over us. She was the current darling of the North American zoo world, another trend I have seen come and go in zoos, and she seemed to know it. She was likely perplexed as to why she was not being courted by us.

Ever polite Jack finally had enough and stopped her in her tracks, telling her in a voice that brooked no argument that the Columbus Zoo would allocate our zoo dollars where we thought best, with the full knowledge that all conservation was risky, but we were willing to take this chance. I watched as the color rose from her neck and spread to her face. Truthfully, I felt a little bad for her, but Jack had shut her down, largely because she wasn't listening and did not recognize that the decision was ours to make. It was a great lesson for me right out of the conservation gate, especially later when I took over the zoo's conservation work: conservation was a competitive business and not for the faint-hearted, but Columbus was willing to take chances and because of that our reputation grew. Supporting field conservation was a gamble, a calculated one, but a gamble nonetheless.

In later years, I would travel with Jack to Zimbabwe, Botswana, and Rwanda. Here is what people need to know about him. He is one of the kindest and most generous people I have ever met. He does not hesitate for a moment when he sees a need. I have witnessed him speaking to a field person who because of a prolonged drought casually mentioned that they were in dire need of a new well and must find the funds to drill for it. Now bear in mind this person was not asking for a donation; he was simply having a thoughtful conversation with Jack about how the drought was adversely affecting local wildlife populations. Jack asked how much money was needed, they replied, and that was it. Jack made sure they got what they needed. Jack also instinctively understood that local people, indigenous peoples, must be involved and reap the benefits of wildlife conservation projects. He knew the two were inextricably intertwined, and one without the other was pointless. Jack understood this decades before many in the zoo world had even the faintest inkling of how integral this approach was.

I have watched Jack continuously facilitate connections between field projects and possible donors and supporters. And that is something I prided myself on at Columbus when I became its first field conservation coordinator. Facilitating connections is probably one of the most effective conservation tools of any zoo. Observing Jack only served to reinforce what I already believed, and he taught me through his example.

8

DISPELLING MYTHS

When I began working with captive gorillas in the early 1980s, numerous myths abounded about them. Gorillas were repeatedly labeled as antisocial, aggressive, would or could not breed, were unable or unwilling to rear their young, and that some females were simply not maternal. One of the most prevalent myths was that females after giving birth did not have enough milk because she showed no obvious breast development—a convenient and frequent excuse for pulling babies. If you were to go back and check local newspapers that ran stories about gorilla births, I would bet this would be a recurrent theme. Whether consciously or not, gorillas were labeled, tagged, and promptly written off. Humans assigned blame to the animals and symbolically dusted off their hands. Problem analyzed, solved, done.

But here's the deal. We, and by that I mean zoos, were talking out of our collective asses. By blaming the gorillas, zoos gave themselves an automatic pass. But the truth of the matter was that zoos needed to take a good long hard look at themselves, own their past mistakes, shoulder their responsibilities, make adjustments, and then alter how they conducted captive gorilla husbandry.

The Columbus Zoo keepers were given a voice and independence, and as such the keepers consciously chose not to pigeonhole those so-called socially misfit gorillas. Our belief was that they were a product of their lives in captivity, not of anything that was naturally inherent in them. I believed, as did my close colleague Charlene Jendry, that given the right circumstances, gorillas would behave as gorillas regardless of their history, background, or past experiences. We recognized their amazing ability to adapt and move on. Several years later, Adele Absi would join us as our fourth gorilla keeper. Her hard work and insightful suggestions would impact our gorilla program as well, often in revolutionary ways.

Mentors come in all shapes and sizes. Some actively function as guides, and others, unintentionally through their compelling stories, serve as a touchstone, a light to guide a person. Bongo was that beacon for me. I was always mindful of his circumstances, of his brutal capture from the wild, his monotonous life in captivity, being on constant public display for twenty-six years, his daughters and son pulled from his mate, Colo, decades before, while he helplessly watched. I can see in my mind's eye his lips pulled tight in anger, his stiff-legged displays, feet stomping in frustration, fists banging against the dividing door as he watched his three infants being removed for hand rearing.

I think about this all the time. His story was not unique; many of these gorillas had suffered similar fates. But there was something about Bongo, maybe his sheer physical presence, his thoughtful all-knowing eyes; it was as if his story and history informed me daily. Often I would catch him in deep thought, his body still, his eyes seemingly far away, as if listening, as if he were calling up some long-ago memory. What was he remembering? His mother, his troop, playing with his siblings in an African forest long ago, the unique smell of his father, his capture from the wild? At times like this, his sadness was palpable, and I moved away before he saw me. I felt like an intruder going through someone's private papers, or bedroom dresser drawers—a witness to something so personal and intimate that I felt chastised. I cannot speak for the other keepers, but for me, all that we would implement, all the husbandry changes, all our initiatives, everything we did over the next fourteen years would continuously loop back to Bongo. All carefully laid at his feet, as if in apology.

In May 2017 a link to a newly published paper on wild gorillas was posted on Facebook. In all fairness, I had not read the entire article but had read a synopsis of it, a somewhat smart-ass commentary that appeared in another scientific publication. The basic premise of the newly published study was that gorilla's "sing" when eating. This was not a new concept to keepers. We have been seeing and hearing these songs for decades so that this was an aha moment for some came as a surprise to me. Gorillas' innate uniqueness and sociability never surprises me, but people's reaction to and interpretation of them always does. These "discoveries" are nothing new to keepers. Just ask them.

Our female Toni was quite a talker or "singer" when having her meals. I have also seen younger gorillas; the twins come to mind, who would sometimes dance about, literally twirling and sing-songing in excitement when seeing certain food items. And, upon receiving their food, they vocalized in a way that I can only interpret as a cross between a relieved sigh and a sense of heightened expectation met. It was a looking-forward-to-eating sound. Think of the level of anticipation you had as a kid on Halloween night, returning home to open your bag of candy, dumping it all on the floor, and then dividing it into "favorite" piles. It's that kind of excitement.

But this particular study focused on wild silverbacks, indicating that they sang more frequently than other group members when eating. The conclusion seemed to be that the dominant male was letting other troop members know he was busy at the moment, was settled in, and was in no particular hurry to move on to other foraging grounds. But I'm a little skeptical because truthfully I imagine troop members are somewhat aware of where their leader is most of the time. And most especially when a male gets up to leave, he usually makes a "naa-humm" or "mmm-waaah" vocalization to give everyone a heads-up, "Hey guys, I'm on the move." My humble theory is something more along these lines. The silverback is simply enjoying his food and is doing what we humans would do when we have a particularly yummy treat in our mouths, eliciting a continuous "hmmm"—a happy humming sound.

9

LEADERSHIP

Jack Hanna's management style back in the early 1980s was unusual then and, to a large degree, is still unusual today. There existed a definite good-old boys mentality among the directors in the 1980s, some of whom had a management style bordering on medieval fiefdoms. In truth, Jack was an innovator, and often when someone marches to his own drumbeat in such a closed, cliquish society, that person is all too often dismissed. Under Jack, not only would Columbus drastically change gorilla husbandry but that innovative atmosphere would eventually lead to the Columbus Zoo becoming a quiet but effective conservation leader beginning in 1990, long before there was pressure to do so. Before supporting fieldwork was considered hip and the right thing to do.

Jack believed in people. Without that belief in and respect for the gorilla keeper staff's knowledge, absolutely none of the positive husbandry practices implemented at the Columbus Zoo would have occurred. Jack taught me what true leadership is: a do-the-right-thing philosophy without thought of how it would benefit you as an individual or whether it was the latest in zoo trends. Leadership was all about making a difference. Jack created an environment that allowed questioning the status quo not in a confrontational manner but rather in a practical and logical fashion. He created an environment where ideas flowed from a more thoughtful perspective and from which innovative ideas were then allowed to be implemented. Jack gave us his trust. He may not have sat in on our brainstorming sessions or heard all the details, that was left to our curator, Don Winstel—but Jack gave us the time, the support, and the environment to truly rethink what zoos were all about. He gave us the gift of a work setting in which ideas built upon one another, in which anything seemed possible, and, most importantly, he created the space to sort things out without the threat of ramifications if mistakes were made.

In 1982, during this fruitful time, the zoo committed to building a new gorilla facility. Jack sent head keeper Dianna Frisch; the zoo's veterinarian, Dr. Harrison (Doc) Gardner; and an architect to John Aspinall's facility, Howletts, in Kent, England. Howletts was a privately owned and operated animal facility that had the most successful captive gorilla breeding and husbandry program in the world. Just as importantly, it had a proven track record of mother rearing of infants within highly social groups.

In the main Ape House built in the 1950s, we had two small outdoor enclosures at the east and west ends, used by Bongo and Colo on the west and Mac on the east. The north gorilla building originally housed elephants but was renovated in the 1970s and housed Oscar and his females. It consisted of two small enclosures in the off-exhibit section with a long linear front area that allowed visitors to view the gorillas through large plate-glass windows. This room connected to a large, grassy outdoor yard with a good-sized wading pool and waterfall surrounded by a deep, dry moat as its perimeter. The outdoor area was much lower than the public viewing area, having a fishbowl feel to it, with the gorillas always being looked down upon. It surely was not a very comfortable situation for them.

In early 1971 John Aspinall completed a large three-dimensional mesh structure for his gorillas, highly unusual in its scope and size. His subsequent success could not be disputed—that is, large family troops and mothers and fathers rearing infants. But it seemed, unfortunately, that many US zoos rejected him as being eccentric or a renegade to the detriment of gorillas in the US population.

Again, it struck me that many North American directors, in their surety, lacked the much needed open-mindedness to (1) recognize the obvious problems facing captive gorillas and (2) be willing to enact the husbandry changes needed to improve their lives. Innovation was suspect, and predictably zoos jumped on whatever trendy bandwagon existed at that time. In this case, "naturalistic" exhibits were all the rage at zoos in the early 1980s. But the reality was that a large swath of flat substrate with green grass did not necessarily make for a good gorilla exhibit. Yes, it was much better than indoor exhibits with only concrete flooring and bars. But think about it: we ask gorillas to live with other group members that they often don't know, and understandably they do not feel comfortable sharing space with on a daily basis. It is similar to being placed in a house with a group of complete strangers and being told to get along. Gorillas are social creatures with complicated likes and dislikes. A big piece of flat land is not enough, not without multiple connected climbing structures (and I mean a jumble of interconnecting structures), not without vegetation to hide in, not without hills, not without ways to get away from another group member. If these very simple elements are not present, it defeats the purpose.

We at Columbus took issue with the "naturalistic" exhibit rage. The real question should have been and continues to be, "What do we need to do to create and provide a complicated, flexible, and comfortable environment so that gorillas can display natural behaviors?" These new naturalistic exhibits were in response to the 1950s barred enclosures, but rather than recognize that three-dimensional exhibits covered in wire mesh might actually work best for gorillas, that concept was thrown out completely because of aesthetics. The thinking was if they were in "cages" they *looked* like they were in captivity, and the viewing public did not want to be reminded of that fact. Once while we were discussing the merits of different exhibit designs, my friend and fellow gorilla keeper Peter Halliday said to me, "The gorillas already get it: they know they are in captivity. Who are we kidding?" The question for me as well as my colleagues was this: who exactly are we building exhibits for—the gorillas or the public?

The Columbus Zoo keepers felt that if we built a structure similar to Howletts we would be providing an environment that allowed the gorillas choices in their daily lives. They could get above the public if they chose—no longer looked down on—or they could choose to be at eye level, observing the public as much as being observed. The point was that it would be their choice. They could escape skirmishes within their troop by going up and over should a fight take place, especially an altercation they may not want to participant in. They could simply choose to go elsewhere. The recognition that choice was the key lay at the heart of our husbandry philosophy. How and why the gorillas responded to any given situation was ultimately up to them. The point was that gorillas made the choice of how to address issues within their own troop. Or if they wanted quiet time away from other troop members, they could go up on a climbing platform or on the overhead catwalk or up to the mesh sitting beds. And if another more dominant member approached them from below with the intent of taking over their coveted spot, they could climb up, down, or sideways along the mesh walls and select another place to sit, thereby avoiding a possible confrontation. It was important that they were never trapped or ever felt they might be trapped. They were given choices by the simple reality of living in a three-dimensional environment. But during this time of planning and gearing up to enact changes, two events occurred that would change and influence our world even further.

IO

AN EXTRAORDINARY YEAR

While preliminary plans were underway to make much-needed changes in the Ape House, our lives continued on, both among the gorillas and keepers. Over in the north building (Oscar's building), there was a pending birth. Bridgette, a wild-caught female on loan from the Henry Doorly Zoo in Omaha was about to make history at the Columbus Zoo. Months before, Dr. Nick Baird, ob-gyn to the gorillas, noticed two heads while looking over the shoulder of fellow obstetrician Larry Stempel, who was conducting the ultrasound. Nick turned to Jack Hanna and said, "Jack, I think we have twins."

Bridgette had given birth numerous times. This would be her fifth, and on the evening of October 23, 1983, while at home playing board games with friends, I received the call that Bridgette was in labor. I was out the door, in the car, and headed to the zoo in a nanosecond. It was decided prior to the birth, that since they were twins, they would be pulled for hand rearing. Mother rearing was not yet front and center at Columbus, but there were definite stirrings of concern about routinely pulling infants. We were at the beginning of turning in a different direction, but we were not there yet. The birth of the twins would actually trigger much-needed discussions about our philosophy: who we were and what we hoped to accomplish. It would also create more changes within our husbandry program. But for the moment, as they were twins, it was thought that it was better to pull them and raise them together for mutual comfort and companionship.

After a fairly quick labor, Bridgette gave birth to the first infant and immediately began caring for it. When the second infant was born, she did not break open the amniotic sac so we moved her quickly to the back holding area. Once Bridgette was transferred, Dr. Baird broke the sac of the second infant. Then Bridgette was lightly sedated and the first infant was removed from her and both infants were immediately taken to the zoo nursery.

What must Oscar, locked in an adjacent back holding enclosure to Bridgette, have felt when he saw the infants taken away? Surely he must have been filled with frustration and rage, a sense of having no power over his daily life, the most basic of rights—protecting and raising his own offspring—usurped. It doesn't take much to imagine how bereft Bridgette would have been. It was another inkling of the great damage we had done to these animals, whether wild caught or born in captivity. Now, many years later, I am still deeply disturbed by the irreparable pain we had inflicted upon the parents and infants.

The public and animal rights people have asked me frequently over the years, "How can you work at a zoo?" My response was this, "Oh, believe me I have plenty of anger for all that has happened to these animals, but I chose to stay in the system, to try to implement change." Did I struggle with it? Absolutely. I was and am the first to criticize zoos for any number of attitudes that were prevalent in the 1980s and for their hubris and lack of humility. I criticized zoological institutions for their inability to even consider, let alone recognize, what they had done, and then to take responsibility for their actions. All of it drove me to do more and to help implement change. And at the core of my work and being, it always came back to Bongo and all that he had endured, all that we had asked of him, and all the grace he has shown throughout his life. In response, we keepers spoke up for the gorillas, we were their voices, we fought for them, and we held others accountable in their names.

Due to the standard breeding loan agreements between zoos, ownership of babies alternated between the dam's zoo (Bridgette belonged to Omaha's Henry Doorly Zoo) and the sire's (Oscar belonged to Columbus). It was agreed as per our breeding loan, that the Henry Doorly Zoo would own the firstborn twin, Baby A, and Columbus the second, Baby B. We also agreed initially that Columbus and Omaha would share in the raising of the infants, each having the twins for five- to six-month intervals, raising them at their respective institution.

The second event that had a profound and lasting effect on all of us in the Ape House was a visit by Dian Fossey. In 1983 Fossey published her wonderful book *Gorillas in the Mist*. That same year she conducted a book tour, giving lectures across the country, including a stopover in Columbus. Having been an avid reader since a small child, I consumed her book. Fellow readers will understand: it was a book that I was reluctant to put down even for a moment. I could not wait to return home at the end of the day to dive back in, devouring every word. It read more like a novel than a nonfiction account of her life and the gorillas. Dian's work inspired and informed many of us working with gorillas. And her eloquent writing made me want to make a difference.

A year or two earlier, I had seen Dian give a talk. The three ladies of primatology, Jane Goodall, Dian Fossey, and Birute Galdikas—handpicked by Louis Leakey to study chimpanzees, gorillas, and orangutans, respectively—were on a lecture tour of the States. I was fortunate to get tickets to see them at a local college. Sitting in the audience that evening, I was inspired by their individual stories and commitment to great apes.

I remember several things about Dian's visit to Columbus, how lithe and tall she was, her soft but husky voice, her penchant for smoking nonstop, and her wry, self-deprecating sense of humor. When Dian came to the Ape House, we had yet to begin renovations, so the enclosures were barren. From the back aisle I watched as Dian walked down to Bongo's enclosure and with bowed head, eyes down, she lowered her body in a submissive manner before him while quietly vocalizing. According to Charlene who was standing very near Dian, Bongo came over with no trace of aggression in his eyes or demeanor, and softly vocalized to her. This, in itself, was extraordinary, as Bongo oftentimes predictably reacted to any stranger in the Ape House with aggressive strut displays.

Dian talked with us about adding bedding to the enclosures and how we might improve their daily lives. Surely, visiting zoos must have been difficult for a woman who had spent years watching gorillas in

FIGURE 10.1. Dianna Frisch, Beth Armstrong, Dian Fossey, and Charlene Jendry
in front of the north yard with Oscar in the tree

the wild, but she was nothing but openly gracious, kindly offering
suggestions and withholding judgment.

I distinctly remember Dian talking about how terrified she was of
giving speeches. After her first public lecture, in which many people
could not hear her because her voice was so soft, a friend came up to
her and shared this advice, "Dian, you have the stories, you have the
knowledge, you hold the key—the audience is simply here to have
you unlock it for them." Many years later when I started giving talks
myself, I would remember the words she shared, and recognize that
the experiences I have had were not mine alone but were the gorillas'
to share with the audience. I was simply the conduit to pass on those
stories. Once you start thinking like that, that it is not about you,
about your shyness, your nerves, it becomes bigger than you and
makes it a bit easier. It doesn't necessarily take away the anxiety, but it
does help to put things in perspective.

In December 2017 I gave a lecture to Columbus Zoo volunteers
about using storytelling as a conservation tool. Up on the screen during
the Q and A was a photograph of Dian Fossey, Charlene, Dianna,

and me standing in front of Oscar's yard with Oscar sitting up in his tree looking on. A former docent and now a zoo employee raised his hand and asked, "Don't you remember that Dian stayed a couple extra days? She thought of you gorilla keepers as some sort of 'earth mothers.'" In an incredulous tone I asked, "Really?" several times. I was surprised and couldn't remember any of that specifically. So, being curious, I stopped by Charlene's house several weeks later and asked her about it. She said, "Yes, remember we were all standing in the Ape House kitchen. Dian was leaning up against the sink, Jack was there too, and she said to him, "I don't know what these women are doing, but they're doing something right." Then I clearly saw it in my mind's eye and heard it, her wispy voice with its unique intonations. What is remarkable about this is that we had not yet even begun renovations or changes to our husbandry program, but Dian must have seen something in us, some passion for change during our many conversations.

During the next year, as the old Ape House was being renovated, the Columbus Zoo began negotiations to obtain more gorillas. Unfortunately, we became involved in attempting to obtain six wild-caught gorillas. I remember talking to head keeper Dianna, voicing my concerns. And even though I was a newbie, in the department for less than a year and a half, I had gnawing misgivings. Something about this whole deal felt wrong, and everything about it sent a flawed message. It encouraged the idea that it was still acceptable to receive gorillas from the wild, regardless of what their particular circumstances were.

For some reason, all of the keepers who would potentially be working with these gorillas had to sign off (actually putting our name to paper) on the application form. I presume because it was the US Fish and Wildlife Service that oversaw importation of animals from the wild. After signing, I had a sinking feeling, a certainty that this was a mistake I would come to regret. That unsettled feeling in the pit of my stomach would be confirmed when I next saw Dian Fossey at an international primate conference in 1985. She gave me the cold shoulder after letting me know what she thought of the deal—and she was absolutely right.

II

CONSTRUCTION

Construction of the twenty-thousand-square-foot Gorilla Villa, or the Habitat as the keepers called it, was completed along with the renovation of the existing south Ape House in 1984, with a grand opening on Memorial Day weekend of that year. Each of the five indoor enclosures was extended using wire mesh on the sides and top, in essence doubling their size; an additional door was added in the lower level of each, and that, combined with the already-existing door on the upper level, created a roundabout system in each. The overhead transfer chutes leading out to the new mesh habitat also had multiple doors allowing us to form roundabouts in them as well. Using mesh was important for the extension of the enclosures as it allowed gorillas to view other gorillas, but the upper section with its original solid walls also allowed them to be out of sight of one another if they chose, just as important as having visual access. The mesh was also a climbing device, allowing them access up and over a situation the gorillas might want to extricate themselves from. We had created a building that allowed them choices, one of the most elemental components in a gorilla's life that had been missing.

Highly unusual was that the indoor facility was closed to the public, an unprecedented decision on the part of management then and even today over thirty years later. Again, this decision was due to Jack. The large outdoor habitat was based on Howletts, and because it was a three-dimensional structure, it acted, in essence, as one big escape

FIGURE 11.1. The gorilla habitat

route for gorillas from their fellow troop members. No one within the group could get trapped or even remotely feel that another troop member might corner them. If a fight were to break out at the north end and started heading toward the south, east, or west side of the exhibit, any gorilla could simple go up and over the skirmish if they chose to. I cannot stress how vitally important this is for both indoor and outdoor living spaces for gorillas. And this type of configuration would allow us to conduct gorilla introductions of all sorts—adult to adult, infant to adult, juvenile to adult—without injuries.

But during the construction of the outdoor enclosure, the gorillas needed to be housed somewhere on zoo grounds. The only option was the old hospital, an antiquated dark holding facility with five side-by-side enclosures and a keeper aisle running along the periphery. Glass windows across the keeper aisle allowed the gorillas to look into the hospital exam room and a few makeshift offices.

In the wild, gorillas build nests on a daily basis. They use some for casual naps in the afternoon as well as constructing more elaborate night nests high in the trees to keep infants, youngsters, and females safe. Large silverback males may choose to nest closer to the ground due to their considerable weight and size. Branches and leaves are bent, torn, stripped, and layered to create a cupped shell. Mothers share their nests with their infants and younger juveniles, creating a cozy snug to sleep in.

Long before wild gorillas were habituated to humans, field biologists studying them used a system of nest counts and counted the number of stools within each nest to quantify the number of gorillas in a specific area. By counting individual nests and noting their size and location, researchers could pretty accurately guess the size of a gorilla troop and determine the number of males, females, and infants/youngsters. Nest building is a normal part of any gorilla's life, a daily occurrence, like eating or sleeping.

Baron Macombo was captured from the wild in 1950. I have often wondered if he was in fact an eastern lowland gorilla rather than a western lowland as he had a very unique face with a sloped plane to it, and his nose print was relatively smooth. He just looked different. Well into his thirties when I first met him, Mac, as we called him, had a sweet nature and was extremely relaxed; he embodied a kind of go-with-the-flow attitude. He lived for decades as all the gorillas did at that time in a small, concrete enclosure, with vertical black bars in the front, a narrow keeper aisle directly in the front of the enclosure, and then a short stone wall with plexiglass windows making up the remaining three-fourths of the façade. This allowed for public viewing—serving as the barrier between the gorillas and zoo visitors. In the 1960s the plexiglass windows were not even in place, so one can only imagine the harsh and unrelenting sounds of the visiting public, which would have assaulted the gorillas on a daily basis. The new windows were installed only after a tuberculosis scare among the gorillas in the early 1960s.

Mac had a television set in the front keeper aisle. Rumors abounded of how he loved to watch football, but truthfully, I never saw any evidence of a preference. He did seem to enjoy watching TV though; looking similar to a kid, chin in hand, sprawled out on the family room's braided rug completely lost in Saturday morning cartoons.

Early on I learned just how laid back Mac was. His eyesight was a bit iffy. He had a tendency to glance at you from the side, showing the whites of his eyes, intently peering as if trying to find just the right angle in order to bring you into focus. And he was very exacting about the order he received his food in the morning, preferring his celery sticks first, before his wedge of lettuce. One morning I made the mistake of doing the opposite, giving him the lettuce first. Quick as can be, Mac grabbed my hand, pulling my entire right arm through the bars. It happened so fast that I forgot all the safety measures I had been told to follow (not that any of the advice sounds even remotely helpful looking back on it now). For example, "If grabbed, drop to the ground, go limp." Absolutely nothing registered. I stood there dumbly stunned and at his mercy. Mac held my arm, while intently and purposely looking into my face, and then he quite simply and gently released me. He could easily have inflicted a bite wound or broken my arm, but he had gotten my attention and made his point. I never made that mistake again.

In 1983, during renovations of the existing Ape House, Mac, along with our other gorillas was housed at the old zoo hospital. There we had the capacity to easily transfer him from one enclosure to another for cleaning. This allowed us to give hay as bedding, which could be removed later during the next day's cleaning routine. One day, shortly after the gorillas' transfer to the animal hospital, Dianna called me on the walkie-talkie, asking me to come down to the hospital. As I entered, I saw Dianna standing in front of Mac's enclosure; she motioned for me to come closer, holding her finger up to her lips, warning me to remain quiet.

Gorillas have a wide array of vocalizations to communicate with one another. In the past Mac might produce the usual greeting vocalization to us but even that was rare. I never heard feeding vocalizations or laughter from him. He was a remarkably silent gorilla.

Dianna had given Mac a bale of hay. A bale contains about sixteen flakes of compressed hay. Mac had shaken out the flakes and was already engrossed in making a nest, completely and utterly oblivious to us. As I drew closer, I heard soft uninterrupted rumbling vocalizations, as if he were having a conversation with himself as he created his nest—a nest of such circular conformity and perfection, with high walls and a cozy deep center, that one would have thought an engineer had designed it. Every few minutes he stepped into the nest to get a feel for it, as if to ascertain the shape and height, and then he'd step back out and make adjustments, adding to, shaking out, and fluffing up his hay, fussing with it continuously—all the while talking to himself. Mac had not had bedding material for nest building since being taken from the forests of Africa thirty years prior as an infant, where he would have shared a nest with his mother.

Mac was transformed, as if in a trance, wholly absorbed in the work at hand, seemingly transported back to his African home. He was doing what all gorillas will do when given leaves, branches, or, in this case, hay. He was making a proper sleeping nest. By the time it was finished, it was truly a masterpiece, perfect in every way. We quietly exited the building, leaving Mac to his memories, both of us touched and once again beguiled by the gorilla's resilient nature.

12

BONGO

Sometimes, I still smell him, a whiff of his sharp pungent odor, like unadulterated human body odor, only cleaner, mixed with a sweet hay-like fragrance. He was stunning, truth be told. When I first began working with Bongo, he was twenty-six, and having reached full sexual maturity, weighed in at just over four hundred pounds. Known as sexual dimorphism, males are twice the size and weight of adult females. His sagittal crest, which supports the development of powerful lower jaw muscles used in chewing tough vegetation, was fully developed, the hair coat on his broad back was completely silvered and short—close to the skin—hence the term silverback for an adult male. As with most adult males, Bongo's hands were humongous with extremely thick fingers ending in black fingernails, hauntingly similar to ours. His face was heavily lined, especially under his eyes, his muzzle round with a down-turned mouth.

I have heard others describe gorillas sometimes as having a puzzled look in their eyes. I understand what they are saying; some gorillas can have a somewhat faraway blank look to their gaze at times, but not Bongo, never once. His eyes were a beautiful rust-brown. They could be contemplative or mischievous (gorillas know when they are being funny), or they might look inward as if he is somewhere else. At those moments I wondered if Bongo was remembering something, especially his childhood, his capture. Those same eyes could also put you in your place with a brief dismissive glance. Bongo demanded respect from us, not intentionally, not because he needed it, but because you knew you were looking at a creature of such grace and wisdom.

I had been working in the Ape House for several months and was constantly brought up short by my good fortune. Gorillas are at once mysterious, complicated, aloof, yet kindly inclusive. One day Bongo decided he wanted to play chase. His signal to play was a "ha" vocalization accompanied by a toss of his head at his back door as I walked past. I stopped, backed up, and peered in. "Yesss?" I said in a drawling goofy voice. With another head toss and an exaggerated "ha," he took off for the adjacent enclosure with another "ha" thrown at me for good measure from the opposite back door when he gets there. As soon as I made my way to that door, off he went again, but this time he did a body roll across the floor before landing perfectly at the original back door I had just left. So we went back and forth, door to door, me in the back keeper aisle and Bongo in his two rooms. By the time we were done, we were both winded, me leaning with my hands on my knees trying to catch my breath, laughing and grinning like an idiot, and Bongo nonchalantly leaning up against the barred door where he threw a self-satisfied, I'm-quite-pleased-with-myself "ha-harrumph" my way. And I again wondered, "How does he find it in his heart to forgive us after all we've done to him?"

It is a cold early spring morning, the cloud cover low and gray, threatening rain as I drive to work. I am about two minutes from the employee parking lot when I notice a dead opossum on the road near the back entrance gate to the zoo. As I drive around it, something catches my eye, a movement. Turning the car around, parking by the side of the road I walk up and see more movement, the opossum mother has several babies that are still alive, crawling on her dead body. I wrap them in a towel, placing them carefully on the seat beside me. I'll figure out what to do with them later but for now I need to get to work.

It is 7 a.m. and our local wildlife rehabilitation center will not be open for several more hours, so I take the opossums into the Ape House with me until I can make arrangements later in the morning to transport them. I am wearing a long-sleeved flannel shirt over my khaki zoo shirt so I button it up, tuck it into my pants, and place the opossums inside my zoo shirt pocket, creating a pouch where they will keep warm from my body heat while I go about my morning feeding and cleaning. The tucked-in flannel shirt will act as a catchall should the babies climb out of the pocket.

Bongo walks over for his breakfast, does a lovely, long "mmmm-wahh" (how are you?) vocalization, and then sits quietly at the front of the enclosure. He is more engaged than usual, noticing straight away something is up; his eyes immediately zoom in on the slight movement inside my shirt. He looks up at me, as if in question, and then makes a gesture with his hand, almost impatient, yet asking at the same time. I can't really explain it but it is comparable to a show-me gesture. Whatever it is, I know exactly what he wants so I place the food bin on the ground then reach into my shirt and pull out a tiny infant opossum, its elongated snout and whiskers work overtime trying to figure out what is going on after leaving the warmth and comfort of my pocket pouch. Opossums are odd-looking creatures at best, strangely cute with teeny needle sharp teeth with a light grayish white body, much darker gray ears, pink feet. Its marble-like black eyes look out from a white face.

Bongo is as intent as any time since I had begun working with him ten months before. He is patient but insistent. I move closer, holding my hand out to him. Bongo reaches out his huge finger, then gently strokes this odd-looking creature. And lets out the loveliest vocalization, a rumbling of approval.

13

BACK TO THE CHILDREN'S ZOO

For two years I had been a permanent part-timer, otherwise known as a keeper trainee, working with Bill and Dianna in their respective areas. Docent Charlene Jendry began volunteering in the gorilla building sometime in 1983 or '84. With the renovations and new habitat, we were gearing up for more gorillas arriving in Columbus and with the influx of these gorillas, I was hopeful that I would be hired full time. I was hosing Oscar's front display room early one morning, when Dianna stopped by to inform me that she was able to get Charlene hired as an additional permanent part-time keeper, simultaneously getting a small pay increase for me. And while I understood the need to have more personnel, it was a shock to me, and I truly felt as if I had been punched in the gut, as if I'd been played. Now that the zoo had two permanent part-time keepers, I realized in an instant that they would not be expending more money to hire a full-time keeper any time soon. I was twenty-seven years old, barely able to make ends meet even with a second job, had no health insurance, no paid holidays, no vacation days or sick leave, so it was with a heavy heart that I applied for a full-time keeper position with benefits, back in the Children's Zoo.

I got the job, and although it was difficult leaving the gorillas, it ended up being a blessing in many ways. But there were many sleepless nights with an absolute sinking feeling of "What did I just do?" So to mitigate that, I spent more time learning about primate behaviors, both through my daily keeper work, but also by exploring less formal ways to increase my knowledge. It would prove to be a wonderful, exhilarating time, a time of great growth and new experiences.

The snow monkeys were still the only large, socially sound, age-diversified group of primates (reflecting what you would see in the wild) at the zoo at that time. These moments spent watching and discussing their behaviors with head keeper Dusty Lombardi were invaluable to me. I also worked a regular shift in the nursery, taking care of the ten-month-old gorilla twins, "the boys" as we called them. They were a delight, rambunctious yet sweetly endearing. A typical day in the nursery started with their morning feed: bottle of milk, bowl of oatmeal, and yogurt. Next I filled the big black rubber tub with warm water, gave them a bath, rubbed baby oil on hands and feet, toweled them dry, then sat with them afterward for a cuddle, burying my face in their still damp fur, breathing in their baby gorilla smell.

They were a handful, acting as a tag team, being typical cheeky youngsters. But if anything happened, if they became frightened, they immediately sought out one another, clinging to each other, or ran to me for a quick leg hug as I patted their backs, talking to them in a soothing voice, and then boom—off they went again.

One day they were being especially difficult, pushing the limits of my patience, when I had had enough. They were constantly banging the public viewing glass, running by me, taking swipes at my legs. There is a certain look young gorillas get—an assessing, thoughtful look with a touch of brash cockiness—a "watch this" look backed up with a self-assured strut. Picking the nearest one, I abruptly grabbed him as he ran by, threw him (gently) on his back, held his arms down by his side—trapping him in essence—then looked directly into his eyes and did a series of short gruff cough-grunts. It's that look a mom gives a kid that says, "I mean business, mister! I'm not dicking around here anymore."

Oh my, did he ever become a wiggle worm, trying desperately to get away, his eyes big and wide in initial surprise, then comically avoiding looking at me—looking every which way than at me. He got the message. And when I released him, he immediately traipsed his shocked little self over to his brother who was waiting quietly in the distance looking very worried indeed. They put their arms around one another, the traumatized twin hooting a tad, his trembling lower lip jutting out slightly. With his brother's arm about him in apparent sympathy, they walked into the adjacent room, commiserating about mean old Beth. They made a pitiful pair, but it made me smile; they had just gotten their first lesson in proper gorilla behavior.

It was around this time that I started talking to Dusty about the need to get the boys out of the nursery and back up with adult gorillas where they would be exposed to the sights, sounds, and smells of their conspecifics. Dusty had had her own epiphany several years before when Cora, who had been raised in the nursery for three years, was then abruptly dropped off one day at the Ape House, having had no previous exposure to gorillas. Her panicked screams were heartbreaking to the nursery keepers as they walked away. Because I had worked with adults, I knew proper gorilla vocalizations and mannerisms. So we discussed using both gorilla vocalizations and a repertoire of gorilla behaviors to teach the twins proper social norms. The sooner they learned the better, and that would make the transition back to the Ape House much easier for them.

There are so many stories about the twins rolling through my head. How sweet they were when tuckered out from playing, finding their way to my lap as they lay down for a nap, my hand resting lightly on their backs. I smoked throughout my twenties, and whenever I came in from a cigarette break, I would sit on the floor, with my back propped up against the wall, legs straight out, and they would gather around me. Young gorillas can look deceptively small, but actually they're quite heavy, as if their bones are much denser than ours; their small size does not correspond to their surprising weight. Climbing onto my lap, their bony rears sharply dug into my legs, as they sat directly facing me, just as they would have done with their mother—especially if she was eating a tasty piece of fruit that they wanted to see and smell in the hopes of her sharing. It's a bit like begging: "Come on, give me just a piece." They both would lean in to smell my cigarette breath, sticking their fingers around and into the corner of my mouth, trying to pry it open. When I finally opened it, I blew gently into their faces. They would close their eyes and then breathe deeply. The look on their faces was very similar to when you blow softly on a baby's face, causing their half-closed eyelids to flutter in pleasure. I don't know what it was about it, but they could not get enough.

14

MORE PRIMATE
EXPERIENCE

Both the north end and south end of the Children's Zoo are beautifully situated along the banks of the Scioto River just north of O'Shaughnessy dam, so no matter where I am I always have a view of this expansive waterway. The honking of Canada geese is constant. In summer they float along the river and browse the grassy lawn near the amphitheater stage. In fall and winter, they fly over in their V formation. The river offers up a consistent breeze in summer, but bitterly cold winds roll off its frozen surface in winter. The west side of the zoo had yet to change into the more themed attraction that it is now. It was simple then, a beautiful piece of land with some open grassy areas, surrounded by large mature trees and the river at its western perimeter.

In summer a local sailing school offered lessons within the safety of the cove near the zoo's aquarium, just tucked off the river. Sometimes I stop my daily hosing and feeding to watch the young kids take to small sailboats, learning the ins and outs of their craft. It was a safe place to learn the ropes, as they were in no danger of going near the dam downstream. Near the sailing cove was Monkey Island, a small stone building surrounded by a shallow moat filled with water. It looked a bit like an ancient slag heap with its uneven set stones. The rest of the all-concrete island was filled with high vertical poles with climbing ropes connecting them. There wasn't a shred of grass or greenery on it for our resident white-handed gibbon family. Gibbons

FIGURE 14.1. Beth on Monkey Island with gibbons

are not monkeys but are lesser apes or, as some prefer, smaller apes.
Gibbons live in monogamous family groups, consisting of a male, female, and their infant. They are territorial in the wild and will define
their home and presence through long, undulating calls heard especially at dawn every day.

To get to the island, I don hip-high waders, place a ladder over
the stone fence barrier down into the moat, and begin the precarious climb down with a cleaning bucket, broom, and scooper in one
hand. I then trudge over to the island to clean the gibbons' small,
fetid indoor area before wading back across with a full refuse bucket,
climbing the ladder to get the buckets of fresh food and water, and
heading back over again.

As with many of the old-school areas at the zoo, the island was
nothing more than a utilitarian structure, built simply for public
display and without regard for the animals' needs. Weirdly enough
though, residents Smiley and Nasty seem to like their summer home,
but perhaps it was understandable as their winter quarters are small

indoor enclosures that won't allow for their acrobatic and exuberant locomotion. The island allows them to swing; they are swift and sure-handed, moving hand over hand at breakneck speed, brachiating, using the interconnected ropes—simulating behavior seen in treetops in the wild. And this exhibit gives them the opportunity to look out over the zoo, the river, survey their territory, and spend the mornings doing their unique long calls. To this day, I can do a mean imitation of a gibbon's territorial call—"whoop whoop whoop woooooo."

The male, Smiley, is tawny colored, with the usual black face framed by a fringe of white. His hands are a lighter shade of gold, and he is slightly smaller than his mate, Nasty. Smiley has a lovely disposition, a real sweetheart, but he tolerates only female keepers and will not allow any male keepers on his island. Oftentimes Smiles, as we call him, comes to greet us, swinging down on his rope system, and then walks in typical gibbon fashion on the ground, his long arms and hands held ludicrously high, walking in a drunken-sailor gait before smoothly jumping up into my arms, asking for a groom or to be held momentarily.

Nasty has a fierce look about her face, with a heavier swath of fur around her neck, her brows furrowed in a perpetual frown. I am not sure how she got her name. Like so many, they come to us with already established names. I like and respect her, as she is a dedicated mate to Smiley and an excellent mother, caring for her infants like a pro. Nasty is known for occasionally swinging down on her ropes and then dropkicking the keepers, especially when Smiley spends too much time with us. She does this to establish her dominance, letting us know who exactly the top female is, and as such, she does not have the time for or need of us, concentrating instead on her family.

In December 1984 I am asked to return to the Ape House to work for several weeks, as the keeper they had hired to replace me abruptly left the department. I am thrilled beyond measure, and it gives me a chance to work with recently arrived gorillas: Donna from the Milwaukee County Zoo, Lulu from Central Park Zoo, and silverback Mumbah from Howletts Wild Animal Park in the United Kingdom. So much is happening in the building with numerous introductions

and breedings with the inevitable pregnancies. And when it is confirmed that Bridgette is pregnant with Bongo's child more than a year
later, I suggest we name it Fossey, after Dian. It is bittersweet to be
back in the building and difficult to leave once again when they hire
another keeper. But sooner than I can imagine I will be back in the
department.

15

A BIGGER WORLD

In June 1985 I asked permission to attend an international primate conference hosted by the San Diego Zoo. I felt so committed to going that I offered to pay all expenses myself. Unbeknownst to me, this conference would have a major impact on me.

While there, I saw a presentation by Wim Mager, founder and director of Apenheul, a remarkable primate facility located in central Holland, where many of the primate groups—spider monkeys, Saki monkeys, Barbary Macaques, woolly monkeys, and marmosets—mix freely with the visiting public, a concept that is unheard of in our litigious culture here in the United States. Wim's presentation included a video of a vacuum extraction procedure used on a mother gorilla that was having a long, difficult labor. After the delivery, they placed the newborn infant on the mother's chest when she wakened, and she promptly cared for it. Apenheul piqued my interest, as it had a commitment to mother rearing that was not yet evident in many North America zoos.

Apenheul had acquired their gorillas from the wild in the early 1970s. Their gorilla troop was led by their male, also named Bongo, and lived on an extremely large tract of land separated from the public by a water moat. The troop shared the island with a group of Patas monkeys. Even the way Apenheul designed their moat was innovative, with a slight muddy decline down to the deepest center—not an immediate drop-off as so many facilities had. The subtle decline reminds the gorillas to take care, that they are getting into deeper water and should be cautious. If an altercation were to happen within the troop and a group member panicked and ran into the moat, they would not fall immediately into deep water, as is the case at many exhibits with moats where drownings or near drownings of captive chimpanzees and gorillas have occurred and continue to occur.

Earlier in my career, I read an article in which an Apenheul keeper carried an infant monkey (that was being hand-raised) into their outdoor gorilla exhibit, which housed a first-time pregnant gorilla. The thought was, by exposing the soon-to-be mom to an infant (in this case a spider monkey), maternal behaviors might be stimulated in the gorilla. The keeper did this on a daily basis and in the end, the gorilla mom, Mouila, raised her daughter, Kriba. What I loved about Apenheul was their willingness to try innovative methods, to go in with their gorillas, not to interfere with them but rather to more closely monitor events.

Several months later in October, I took my first trip to Europe. Armed with a Eurail pass, I landed in London, took the train to Dover, boarded the ferry to Ostende, Belgium, and then took a train north to Amsterdam. A friend was also in Holland at the time, visiting his family in Amsterdam. Harold had worked at the Columbus Zoo before moving on to the San Diego Wild Animal Park. It is one of those fortunate events that through this friendship unexpected doors opened to me. While visiting Harold, we dropped by one of his friend's in the heart of Amsterdam, along the Albert Cuypstraat. We met his friend's brother, Robert, and as luck would have it, Robert's girlfriend, Joke (pronounced Yo-ka). Joke lived north of the landmark Central Station. She was leaving to go home to feed her cat and casually asked if I wanted to tag along. Hopping in her car, we headed to her place. As we drove, she mentioned she was going to be living in Paris for the next year, working as an au pair. Through our conversations I learned that Joke was coming to the States in November, so I invited her to stay at my place. The pieces of the puzzle were being laid out for me to put together.

Several other events happened simultaneously during the late fall of 1985: a full-time keeper position was going to open up at the Ape House in early 1986. And although I was going to apply for it, I knew in my heart it was a done deal, that my friend and colleague, Charlene, would get the job. So the question for me was, what could I do in the interim that would be a positive experience and increase my primate knowledge while at the same time get me out of the zoo for a while?

When I got back from Europe, thoughts about taking a leave in order to visit Apenheul to study their gorilla program began swirling around in my head. So when Joke visited in November, I asked "Would you be willing to rent your apartment?" This had been a huge piece of the puzzle, a place for me to stay while in Holland. When Joke said yes, it was as if it was meant to be. Now, how to actually get permission to go to Apenheul? With that one question, my excitement about the possibilities began in earnest.

I approached Children's Zoo head keeper Dusty about taking a leave of absence in order to visit and informally study at Apenheul. And with her complete support, I then talked to Jack, who at that time was still very much involved in the daily running of the zoo and always had an open-door policy with his staff. I stopped by his office, asked his assistant if he was busy, and she waved me through. After a brief conversation with him, I agreed to put something in writing for everyone so it would be on record what I hoped to accomplish, and I was given permission to leave for two months. The deal was I would take a leave without pay, would pay all my own expenses—including flight, lodging, food, and in-country travel—and the zoo would hire a part-timer to replace me for those two months. I will say this to anyone: if you want more experience, more knowledge, be willing to pay for it; it is worth it. I also will say this: it was no small sacrifice for my fellow keepers who covered for me during my absence and for that I am forever grateful.

I am sick of it, the waiting, my nerves frayed. I spend sleepless nights worrying about everything, about being in a strange country, about getting to Apenheul from Amsterdam, about navigating a new city. Ten days out from leaving, I am now nothing but impatient and on edge. I have done all the prep work. I've gotten permission from my zoo, written to Apenheul founder and director Wim Mager, asking if I can come, have received his formal letter back giving me permission, made copies for Dusty, Don Winstel and Jack, saved my money, and arranged for a friend to take over my second job for two months.

I have booked a flight and paid the rent for my two-month stay. Joke has been kind enough to arrange to have Robert pick me up from Schiphol airport and take me to her apartment. I am beyond ready.

Grubby from my long, overnight flight, I breathe a sigh of relief once I see Robert's familiar face. Twenty-five minutes later we pull up to Joke's apartment. We unload my bags and carry them up the steep set of stairs to the living room. Robert shows me how to light the gas heater stove—the only source of heat for the two-story apartment. My home for the next two months is a charming space. It has a small, linear kitchen with stove, sink, and white cupboards on one side and a door to the combo shower/toilet room directly opposite. It has a living room with two small sofas, a coffee table, side table, and small TV. The room is light and airy, lit by several large windows. It is cozy or, as the Dutch say, *gezelig*. Up another short flight of stairs, there are two small bedrooms.

Over the next couple of weeks, I take the time to acquaint myself with Amsterdam. It is a fifteen-minute walk from Joke's place to the ferry landing, where I board the ferry that takes me the short ride across the IJ River to the back of Central Station. A quick walk through the train station brings me out into the central part of Amsterdam. I walk the streets, stopping in cafés, eating wonderfully simple meals of soups and sandwiches made of crusty bread with thin slices of ham and cheese. Every day I buy the *International Herald Tribune* and find a café, where I sit, drink coffee, and read about what is happening in the world from a distinctly European perspective. There are nonstop daily reports of the space shuttle Challenger explosion, which my fellow keepers and I had watched in the Children's Zoo staff meeting room just a few short weeks before my departure. My family lives on Merritt Island where Kennedy Space Center (KSC) is located. People walk out their front doors to see rockets and space shuttles blast off. Many work at the space center or work for contractors at KSC. My younger brother, Carl, watched the launch that day and knew immediately when the smoke pattern changed that something catastrophic had occurred. He has never watched another launch since.

I become familiar with the concentric canals of Amsterdam's city center, the Herengracht, Keizergracht, and Prinsengracht. I meet Robert for drinks. Robert's mother invites me for dinner one evening and I arrive, flowers in hand, for an evening of stimulating conversation, with classical music playing in the background. It is a simple but typically elegant Dutch meal of potato leek soup, bread, an entrée, and dessert. Mrs. Strada (Nell) and I plan a day at the movies, meeting to go see *Out of Africa* at the Leidseplein on the outskirts of the city. Afterward we go to the American Café with its art deco architecture, stylized hanging chandeliers, and dark mahogany wood tables. It's beautiful and sophisticated and seems as if from another time. We drink dark rich coffee accompanied by delicious flaky pastries and discuss our impressions of the movie.

I get into a semblance of a routine, shopping at the local grocery store, sometimes eating at a little café near the river crossing. I hand wash my clothes, using the spinner to drain and semi-dry them before hanging them out the kitchen window. I learn to say *dag* and, more formally, *goede dag* (good day) with its guttural *g*, *alstublieft* (please or here you go), and *dank u wel* (thank you). I can order my coffee and tea in Dutch, even my hot chocolate *met slagroom* (with whipped cream).

As it is an exceptionally cold winter in 1986, all the canals freeze. One night Robert takes me to a lake where everyone is skating. But this is not like my childhood, skating in white figure skates. Instead I am skating on long-edged speed skates—the national pastime in Holland. I feel as if I am flying, the cold biting my cheeks as I get the hang of it, the rhythm of skating low, my arms swinging by my sides, feeling joyful holding my balance going around corners. I love it. I am a kid again on my neighborhood pond.

16

APENHEUL

On my first train ride to Apenheul, unbeknownst to me, I neglected to change trains in Amersfoort, about twenty-five minutes outside of Amsterdam. After several hours, I end up in northern Holland, and when the conductor comes to get the ticket, he realizes what I have done and says something in Dutch. I am stricken, then embarrassed, and then mortified when I look up to see the other passengers turn in their seats peering at me, some with half grins on their faces. The conductor takes pity on me, gets me on the right train back to Amsterdam, with no charge. When I finally arrive back at Central Station, I head directly to the nearest bar and get properly toasted.

But several days later I finally make my way to Apenheul where my primate education begins. Each day, I arrive with notebook in hand and conduct an informal study of their gorilla troop. The troop consists of adult male Bongo, adult female Mouila, her seven-year-old daughter Kriba (the mother/daughter gorillas from the article I had read years before), Mandji with her four-year-old son, Lukas, and Mintha with her two kids, five-year-old Frala and two-year-old Dihi. Also in the troop are two additional adult females, Kim and Tsimi, who have so far not reproduced. Another female is Lobo; she has an eight-year-old son, Dibo, whose biological father was shipped off to another European zoo prior to Dibo's birth. Shortly after his birth, the resident silverback Bongo attacked Dibo and injured him quite severely. As a result, Dibo's head is slightly misshapen, and he tends to drag an arm and leg, especially when tired or worn out. He often has a slightly lost look on his face and is a little slow physically and mentally.

It's important to remember that Columbus did not yet have a "normal" group of gorillas, so watching these gorillas interact was like an addict getting a daily hit. I simply couldn't get enough. Everything I saw either confirmed what I already believed or knew to be true or was a revelation. It was sheer pleasure to observe them each day.

My Apenheul observations:

In an upsetting situation Frala and Dihi will seek out their mother (Mintha) for reassurance as will Dibo with Lobo, and Lukas with Mandji. All juveniles/infants share a night nest with their respective mothers. Kriba, although seven years old, will often sit with or near her mother, Mouila.

Bongo appears to be a tolerant father and often plays with his offspring:
11:00 a.m.—Lukas chasing Bongo around entire exhibit.
11:08 a.m.—Lukas and Frala both chasing Bongo around exhibit.
11:18 a.m.—Bongo, Frala, Lukas, and Dihi playing together.
11:26 a.m.—Frala starts at opposite side of exhibit and starts running top-speed towards Bongo. Bongo runs around exhibit twice with Frala in tow.

My casual conversations with the keepers taught me much about their husbandry: "When keepers feel that a birth is imminent, the pregnant female is separated and housed with her 'best friend' from the troop—in this case Manji. Within twenty-four hours of the birth, an introduction back to the group has occurred."

Each morning I worked in the gorilla building preparing diets, cleaning enclosures, and helping around the building. I took a mid-morning tea break with the staff, asking questions and raiding books from their extensive library. As in any new situation, it was awkward for me at first, but each day it got easier as I started to get into a groove. The gorilla keepers, most especially head keeper Frans Keizer, took me under their wing, patiently answering all my questions. After only a few days, they asked me to feed the gorillas. This would have been unheard of at Columbus. Our gorillas became agitated when strangers come through the building, but at Apenheul, it was more relaxed, more *laissez-faire,* the gorillas more comfortable. It was not unusual to see several dogs, one the size of a Saint Bernard, wandering into the gorilla building, meandering down the back aisles. The gorillas rolled with it, and when I fed them there was only one brief cough-grunt from a young female. They didn't seem to mind me in the least. Weeks later, when a keeper calls in sick, I am asked to take care of the Barbary macaques, squirrel and saki monkeys, and cotton-top marmosets. I am gratified that the staff trusts me and treats me in such a matter-of-fact manner.

Every day is a learning experience. I ask questions about the design elements of the building and Apenheul's diets and husbandry philosophy. The building is made of numerous interconnecting rooms that can be shut down into individual spaces, thus allowing the staff to separate animals for feeding, observations, and births. Each room has a large plate-glass window. Visitors enter the square of interconnecting rooms by walking under the overhead glass chute, then stand in the middle of the unheated center of the exhibit—in essence they are surrounded by gorillas who are on all sides.

Because of the huge windows, all the gorillas are able to see what is going on in the troop regardless of which section of the exhibit they happen to be in. In essence, it is a roundabout—albeit a square one. The windows allow a gorilla in the west end of the building to peer over and through the visiting public to the east side of the exhibit. If an altercation breaks out, that visibility allows each gorilla to choose how they want to react, to either get out of the way if they see a tussle heading in their direction or actually go to the aid of a fellow troop member involved in the scuffle. The building is not pretty, but its fundamentally simple design is absolutely brilliant because it works for gorillas. Years later, the Columbus Zoo would build an additional gorilla exhibit in the late 1990s; it was the Apenheul design we followed. Interconnecting rooms and chutes created one big roundabout—with the public in the middle surrounded by gorillas on all sides.

At Apenheul, two doors lead to the outside yard so no one gorilla could be blocked from either entering or exiting the building. This is an essential element because the gorillas are given access outside even in colder weather. The gorillas are acclimated to cooler temperatures because they kept the building at 51 degrees (11 Celsius).

The gorillas' diet is diverse and includes fruits, vegetables, lots of daily browse (tree limbs), and a specially made protein ball of oats, molasses, protein powder, and vitamins. I copy the recipe and practice making it, intending to take the idea back to Columbus. Every afternoon, the gorillas are given tree branches and the adults spend hours stripping the bark while their kids play or nap. When I clean the next day, I am amazed at the bare white bones of the branches where the bark has been completely stripped. I also notice that the gorillas' stools are consistently segmented and formed. Rarely do I see loose stool, which I can only attribute to their diets and the daily browse in particular.

Curator Tine Griede is kind, offering to host me during the week. So rather than traveling to and from Amsterdam every day, I now come on Monday and stay through Friday, returning to Amsterdam for the weekend. Tine's offer is lovely, and we get into a routine of cooking in the evening and sitting out back in her small garden with a drink before dinner. I not only spend time with Tine but with her sweet, longhaired dachshund, Hobbelien, who accompanies Tine to work every day.

It wasn't all work. One day Tine had to go to Arnhem to the Burgers' Zoo where Frans de Waal did his groundbreaking observations on their group of chimpanzees. I was somewhat familiar with de Waal's work, having seen his presentation at the 1985 primate conference in San Diego. If I remember correctly, he showed a fascinating video of a chimpanzee systematically experimenting with different-size tree branches, propping them against a fence (or a tree?) to use as a climbing bridge of sorts, which eventually allowed the chimp to escape the exhibit. I was thrilled to be asked to join Tine on her visit. The Burgers' Zoo housed the six wild-caught gorillas that the Columbus Zoo had attempted to obtain years before. It's odd how things work out, how things loop back around.

Another weekend, rather than going back to Amsterdam, Tine and I took the train to Paris to visit Joke. After our arrival, we immediately checked into a classic old-school Parisian hotel. Our room was charming, with two narrow beds, high ceilings with elaborate crown molding; sheer white curtains graced the ceiling-to-floor win-

FIGURE 16.1. Beth with woolly monkey and Tine's dog, Hobbelien, at Apenheul.
Photo credit: Beth Armstrong.

dows. We decided to take a quick nap before meeting Joke. I lay
there, marveling at my good fortune to be in Europe, to be in Paris.
A soft breeze gently blew the diaphanous curtains back and forth in
an elegant dance, and in the distance I heard the distinct clicking of a
woman's high heels as she walked along the cobblestone street below
our windows. It was dreamlike and magical, as if out of a movie.

It wasn't just gorillas I was learning so much about at Apenheul;
my love affair with woolly monkeys began there. Woolly monkeys
are beguiling; they are grayish dark brown primates with rust-colored
eyes, prehensile tails, and hands so eerily human-like. The woolly
monkey troop at Apenheul is allowed to wander freely with the public, minus the adult male, as he could be aggressive when protecting his females. Woollies have a coat of deep dense fur so uniform
in length and fullness that it looks similar to a closely cropped sheep
groomed to perfection and ready for show at the state fair. Their faces

are enchanting—some have a protruding lower lip and all have furrowed brows that make them look as if in deep thought.

One morning while I sat on a bench outside the gorilla building, a woolly sauntered over to sit at the opposite end. The keepers have schooled me in the fine art of a woolly "snuffle." Hand over mouth, my head tilted down and away in a coquettish manner, I averted my eyes while making a slurpy in-and-out sound. It's an invitation to "come and let's have a good snuffle together." This woolly, as with most, simply can't resist, so she sits next to me and off we go, mimicking one another's every move and vocalization. As we happily snuffle away in unison, I am charmed beyond all measure.

17

BACK TO GORILLAS

Upon my return to Columbus in early April, I found a new place to live, a lovely old-school apartment in my childhood neighborhood. The brick buildings are lightly coated in white paint, offering a glimpse of the red brick beneath. The colonial-style apartments are adorned with green shutters and graceful arched walkways. Heavily wooded ravines are scattered throughout, and mature oak and sycamore trees grace the many courtyards. Built in the 1920s, the complex has an Olympic-size swimming pool half a century old and a community garden for the residents down by the Olentangy River. It is home to an eclectic group of residents: college professors, long-time elderly residents, graduate students, and some young families. I'm sure its simple beauty drew many of us. It is a completely delightful place to live.

Sooner than I had dared to hope, another full-time position opened up in the Ape House, and I am hired back, starting in June. I am over the moon to be working with gorillas again. Bridgette's pregnancy, which was confirmed earlier in the year, continues through the spring and summer months without any problems. After the birth of twins in 1983, Bridgette had an inordinate amount of bleeding, so Charlene and Dianna researched and implemented natural dietary supplements into the gorillas' diet. They established a daily routine of giving raspberry leaf and alfalfa tea to pregnant females, a uterine tonic that helped diminish blood loss after giving birth. Fennel seed, fenugreek, and oatmeal water were given to promote milk production. We also used herbal teas for other purposes. If a gorilla was nervous, we gave chamomile tea twice a day. When a gorilla showed signs of coming down with a cold or flu, we made a big pot of onion garlic soup to give twice daily, as well as extra vitamin C tablets. Could we prove that any of these natural remedies worked? Nope, but we believed they did—and the zoo believed in us, allowing us to explore this approach.

It is not unusual nowadays to see animal areas adorned with all manner of enrichment items at zoos: cardboard boxes, garlands of colored paper, burlap sacks, and treat stations. Decades ago enrichment as a husbandry tool was still in its infancy, but for a number of years Dianna had been brilliant at coming up with unique enrichment items long before it became the norm for captive animals. She called local fire departments asking them to donate their discarded fire hoses. Sometime we used the lighter, narrower hoses, but as they are more flexible, they can easily be looped and twisted, making a noose, an obvious hazard to young gorillas. But Dianna was creative and found a use, weaving them into sturdy sleeping mats, which served as hanging beds.

The wider gauge fire hoses could be used as a single strand secured at each end from the mesh ceilings. Silverback Mumbah loved to use these as hammocks. He carefully lifted his bulk into the loop, lying on his back and stretching his legs comfortably up one side of the hose while his head rested on the opposite side. He swung ever so slightly from side to side while remaining perfectly balanced. It wasn't unusual to see the twins sitting on either side of Mumbah, reaching out and gently pushing him back and forth.

Dianna learned that telephone books were printed with nontoxic soy ink. So a call was sent out to zoo docents and others to donate their discarded copies. We got loads of them, tore off the covers and spines, and then drizzled honey or yogurt on the yellow pages, using them as distractions during our numerous gorilla introductions, randomly distributing the pages throughout the enclosures. Sometimes I would walk by and glance in only to see a gorilla sitting nonchalantly holding a phonebook, sifting through its pages, looking for all the

FIGURE 17.1. Mumbah on rope swag in the habitat

world like she was trying to find a phone number when in truth she was searching for honey-soaked pages.

A Wendy's restaurant was built directly behind the Ape House. Being a good neighbor, the manager stopped by to introduce himself. While giving him a tour around the building, he offered to help in any way he could. So when we had difficulty getting the gorillas to take a particularly nasty medication, we ran over to Wendy's for pure coke syrup, as its sharp tangy flavor could mask anything. When we housed a pair of young orangutans, Teak and Amber, who frequently and stubbornly refused to come inside in the evening, looking at us with their "sorry, no can do" baleful eyes, we headed over to Wendy's for a chocolate Frosty. Offering them spoonfuls of their favorite treat, we moved them step by step through their transfer chute to their indoor sleeping quarters.

The manager offered to build us a few treat stations, and they were a big hit in the Ape House with both gorillas and keepers and are still in use today. Made of a two-by-four wooden board set at a slightly sloping angle, three holes were cut out in the diameter of a plastic Frosty cup. "Dipping" cups were then placed securely into the holes.

The upright stand is about two and half feet tall, also made of two by fours. We anchored the treat stations in the drainage well in front of the indoor areas, filled the cups with yogurt, peanut butter, honey, or sometimes even a Frosty, and gave the gorillas tree branches.

Using a tree branch as a fishing tool, the gorillas threaded it though the mesh, dipping into the cups and then pulling it back in. With yogurt or honey running down the length of the branch, the gorillas licked it off, looking similar to a kid eating a melting ice cream cone on a hot summer day. Some gorillas pulled back too quickly, desperate to catch the drips. But it's the really good toolmakers that reap the most rewards. They bother to cleanly strip the branch of all leaves and small protruding twigs and then break the fishing branch into just the perfect length. When they thread the honey- or yogurt-dipped end back through the mesh, they are very careful not to touch the mesh. It looks a bit like playing the board game Operation, when a kid carefully tries not to set the alarm off by touching the sides. Once the branch is clear of the mesh, the gorillas quickly slip a hand under the end of the stick to catch any drippings.

Colo was exceptionally good at fishing for the treats. When she was busy at the dipping stand, one of her adopted youngsters, either JJ or the twins, would sidle up next to her. They intently minded her every move, watching her take the branch to her mouth, their faces only inches away from hers, their disappointment clearly showing as she stared doggedly straight ahead, never once glancing in their direction. Although she was incredibly tolerant and patient with the young ones in the troop, on this one point—dipping—I'm not sure she was so inclined to be magnanimous. Their woeful boo-hoo expressions never moved her. She rarely shared the fruit of her meticulous labor.

When I returned from Apenheul armed with my copious notes, I started making suggestions. We adjusted our gorilla diet to better reflect theirs. My observations at Apenheul suggested that we should provide browse more frequently—daily if possible—and incorporate something similar to their protein balls into the gorillas' diets. Dianna had initiated giving browse years earlier, but we had yet to establish

FIGURE 17.2. Adele Absi (Dodge) giving rose petals to gorillas as enrichment

a system of getting browse consistently. Dianna took the reins and got it going. And rather than protein balls, Charlene and Dianna came up with a recipe for protein drinks that were given three times a week. Their drinks were given in an array of large colorful plastic cups—red, yellow, pink, and blue—with each gorilla's name neatly written in black marker.

Having seen climbing structures in the indoor areas at Apenheul, I suggested we construct some in each of our indoor enclosures, and we did. Some were more elaborate with side-by-side wooden slats, making beds for resting and sleeping, while others were simpler in design with long beams of wood similar to balance beams raised several feet off the floor. Apenheul taught me to look at the existing space and see the emptiness that needed to be filled. Filling in that dead space became a priority for us, allowing the gorillas varied options on how to navigate their enclosures. This way of looking at an exhibit, whether indoors or outside, forced us to examine the space from the gorillas' perspective; we stood in the exhibit and actually asked questions like, "Would I feel trapped here or here or over there?" "Where

FIGURE 17.3. Gorillas sitting on wooden platforms built in the habitat

is my escape route from anther troop member—how do I get away?"
Our mantra became "How would I feel?"

In the summer of 1986, we flew Tine Griede over from Apenheul
to give a lecture at the Columbus Zoo. While here, we picked her
brain about our exhibit design and husbandry practices. She immediately suggested that we build some hills in the habitat and even more
climbing structures throughout our enclosures, both in and out.
Thanks to our Columbus management, within a very short time after
her visit we had a hill with a tunnel through it, which the youngsters
loved. We added more elaborate climbing platform structures to the
habitat. Those changes thirty years ago are still in place today and are
being used by the great-grandchildren of the resident gorillas from
back then.

We spoke with our maintenance crew, explaining what we needed
and what we were trying to accomplish. They took it to another
level by creating simple but ingenious metal foot- and handholds in
the shape of an elongated half rectangle—each end securely bolted
to the cinder block walls. Once again, these additions had multiple

purposes. They gave gorillas an exit route up and away from a fellow troop member, while making it easier for all gorillas, large and small, to navigate their living quarters, to climb up into chutes and onto sleeping beds. The maintenance guys also built and installed triangle-shaped corner beds made of metal mesh that, when bedded down with hay, were quite comfortable. They even mounted a couple of corner bunk beds in several enclosures—one on top of the other. We slowly but surely began to transform the gorillas' living areas. The once long empty shells were now being made into a vital part of their everyday lives.

There is nothing better in the afternoon—especially on an unhurried winter day right after the afternoon feeding, when the building is quiet and homey and when fresh hay has just been added—than to sit down for some observation time.

The gorillas love bedding. They scoop up mounds of hay in their arms, duck-walk over to their preferred sleeping bed, and use the hand- and footholds to clamber up the walls to a higher elevation. Some may lose half their hay load in the process, so they stoically climb back down to retrieve lost bedding and then scramble back up. Getting perfectly situated in their bed, they begin the process of shredding and shaping, creating a large comfy nest—a bit here, a bit there until satisfied, and finally settling down on their side, their legs curled up into their chest, with their hands cupped together under one cheek for a pillow. When they spy me watching them, they let out a friendly "mmm-wahhh." After years of observing gorillas with their similar mannerisms and behaviors, I am always struck by how unnervingly alike they are to us.

Dianna researched wood wool as an alternative bedding material. Wood wool is cream-colored, finely shaved slivers of wood; it's very delicate but when packed together in a bale is quite dense in weight. It looks similar to the type of material that lines a gift basket of cheeses and sausages you might give or receive at Christmas. We ended up using both hay and wood wool, switching back and forth for added stimulus. Wood wool is much more expensive than hay but is easier to clean, as it is obvious what is soiled and what is not—allowing us to only remove the soiled bits daily. Some gorillas preferred hay, some wood wool. Interestingly, the ones that made large elaborate nests didn't seem to have a preference. They used what was available, consistently making fantastic nests.

I had become enamored with classical music, Mozart's *Eine Kleine Nachtmusik* and Vivaldi's *Four Seasons* in particular. We had a tape deck in the front aisle, and each morning after we made our rounds, we would slip in a tape of Mozart, Vivaldi, or my new favorite, Bach's Brandenburg concertos. Then we would begin clearing soiled hay, hose down floors, apply a light disinfectant, rinse, and finally squeegee the excess water. We finished by using industrial-sized fans to help the floors dry more quickly, but inevitably the floors were still damp when we let gorillas back together. Our silverback, Mumbah, absolutely hated getting his feet wet and would predictably come down from the chute with his toes gripping two generous pieces of clean wood wool. Shuffling his feet along the damp floors, he looked very much like he was wearing big fluffy slippers. All he needed was a bathrobe and morning paper tucked under his arm to complete the look.

And Mumbah was a funny one, somewhat finicky about certain things. Building Mumbah's troop, as we added more and more members, coincided with increases in our enrichment program. We added small metal pans, large plastic trashcan lids, coconuts, blankets, and burlap sacks. The twins loved to place their hands on the trashcan lids, scooting across the enclosures, pushing the lids in front of them, a bit like skating. With apparent satisfaction, the boys frequently dropped the metal pans from the highest heights, and the *clang, clang,* and *clangs* resounded through the Ape House. The hard husk of coconuts made the most amazing organic sound when dropped, pushed, or scooted. Think of Monty Python using the husks to mimic the clopping sound of cantering horses. When the twins were being raised in the gorilla nursery, one of their favorite games was to hide under big plastic trashcans and then randomly scoot around their exhibit. They were like children draped in sheets playing ghost blindly bumping into one another and laughing before peeling off their covering and beginning another game of chase.

Coconut husks and the smallest of the metal pans were frequently used by Colo as hats. The enrichment items scattered throughout were a veritable millinery for her to choose from. She walked carefully, balancing her chosen headgear, like a preteen girl carrying a book on her head in order to improve her posture. She lifted a hand only to right her headgear if it started to slide, firmly securing it before continuing on.

Mumbah, on the other hand, seemed to find all of it utterly annoying. At the start of every morning, we would find a pile of pans, husks, and lids outside the back enclosure door where they had been methodically collected and pushed out, dropping down to the keeper floor below. During the day, if the boys became too loud and boisterous, Mumbah quietly but purposefully walked himself over to the offending youngster, and, if they hadn't already dropped the item, he would gently but firmly take it, walk to the back door, and push it out. Eventually he took to doing the same with burlap sacks and blankets, so maybe it wasn't the noise but the clutter that bugged him?

18

TONI-BALONEY

The general consensus has always been that silverback males are the dominant force in any gorilla troop, but some of us who have worked with gorillas may have a slightly different take. Females, who are at once sophisticated and shrewd in their subtle behaviors, can quietly be in charge in any given situation if they have a mind to. Dominance can also be as simple as a friendship, a preference resulting in an alliance.

In 1985, amidst the bustle of building the Columbus Zoo's gorilla program, we received male and female gorillas from all over the country. Sunshine, a blackback, came to us from the San Francisco Zoo at the age of eleven. He was sexually mature but still juvenile in behavior and to some degree in looks with no sagittal crest to speak of. He was long limbed, and as awkward as a teenage boy who had sprouted twelve inches in a year, looking like some strange cross between ancient man and an ape species.

Our female Toni was born in 1971 at the Columbus Zoo, the last of Colo and Bongo's offspring. She had been raised in the nursery and spent much of her adolescence in the company of her sister, Emmy, and their brother, Oscar, whom she eventually bred with, producing several infants (don't ask—it was a different time, an unenlightened time). Toni was a skittish gorilla and made a conscious effort to avoid any and all confrontations within any gorilla group she was in. She also has an arsenal of odd ungorilla-like behaviors, with walking

FIGURE 18.1. Toni and Sunshine playing

upright an inordinate amount of time being the most apparent. Another was her daily habit of fluttering her hands neurotically around her mouth while standing upright, facing the wall. Toni-Baloney, as we sometimes called her, was fixated on jewelry, most especially large watches. While being raised in the nursery, she had been taught some cursory sign language. She would often sit quietly obsessing over a keeper or visitor's watch, bracelets, or rings, tapping her wrist where an imaginary watch might have been and then point to her mouth to indicate that it was time to eat. In some ways, it seemed that Toni never truly reconciled to being a gorilla. Always a bit mentally removed from the other troop members, Toni lived in her own world. Later in life a change would come when she became a surrogate mother to two infants—proving as usual to never label gorillas.

But Sunshine was the gorilla she was most comfortable with and it wasn't unusual, while we were working in the kitchen or having a cup of coffee, to hear continuous shared laughter. We would look at one another with raised eyebrows then quietly rise from our chairs to stealthily make our way down the back aisle to sneak a peek around the corner to glimpse the two of them in the throes of a tickling bout. When they realized they had an audience, they stopped and looked a bit sheepish as if caught in an illicit act, but then, more times than not, they seemingly shrugged their shoulders as if to say "fuck it" and then got on with their play.

Toni became the dominant female in Sunshine's troop not because she actively sought it but because Sunny seemed to prefer her company. He simply liked her more than the other females. Unlike most females intent on moving up the social ladder, Toni never used his preference to her advantage.

Some adolescent males—who have yet to acquire the wisdom, judgment, and restraint to woo their females into doing what they want—make brute force the order of the day. Eventually males grow out of this, but it can take years before they recognize the finer points of nuanced persuasion. Sunshine was no exception and in some ways was even more extreme. He could be quite rough with his females, most especially immediately after copulating, administering a swift kick or bite to the female after the act was consummated. The keeper staff found a solution by plying him with a bit of Mad Dog 20/20 prior to breeding, which seemed to take the edge off, much to the relief of his somewhat reluctant and nervous mates.

During Sunshine's time of maturation, Toni and the other females put up with his over-the-top roughness. He was insistent in play and even normal social interactions to the point that he first annoyed the females and then created uneasiness, resulting in the targeted female just wishing to get away from him. Even though he and Toni were the best of friends, she was not necessarily immune from his immaturity.

One day Sunshine kept bugging Toni, nothing in particular, just a sort of relentless irritant, like a kid poking a finger in another kid's arm, just to get attention. Sunshine and Toni were in an indoor area, which had a cement sleeping bed tucked into the corner, about four feet off the ground. Sunshine is the tallest gorilla I have ever seen in captivity; he is lean with extremely long limbs. To get into any small spaces he has to collapse his body, in essence folding into himself.

That day, Sunshine kept at Toni until she finally snapped. She screamed at him across the length of the enclosure, backed him into and under the elevated corner bed, and kept him there with her wrath. There is nothing quite like the sound of a pissed-off female gorilla shrieking full-throttle. It is startling for the uninitiated; you would swear that someone was being murdered, or about to be.

Sunny cowered under the bench, holding his hands near his face in a defensive posture, his eyes blinking in alarm. But once Toni backed him into the corner she seemed satisfied; point made, as it were. Turning her back to him, she proceeded to walk away in her usual upright position, potbelly sticking out, back arched, a purposeful dignity coupled with a rather silly walk.

Sunshine, thinking the coast was clear, cautiously began to come out from under the bench. Big mistake. Toni with her back still to him paused for a moment, pondering as if remembering some long-ago slight, before abruptly turning around to him as if to say, "Annn-nnd another thing, Buster!" She screamed him right back under the bench where he tried to make himself as small and invisible as possible. Content that her point had been made (again), she moved on in a self-righteous kind of way. Sunshine rather intelligently waited a while before he quietly extricated himself, and they then both went on about their day as if nothing unusual had transpired.

19

MOMENTS OF MAGIC

Doc is absolutely furious. Rail thin with a commonsense, humble approach, Dr. Harrison Gardner is our zoo veterinarian. He listens well to us, the keepers, and rarely raises his voice. He is a man of few words, and we have the highest respect for him and his quiet presence. Usually he has a wonderful smile that carries all the way to his eyes, but not today. Today I am in a zoo van holding a baby gorilla that we have pulled from its mother, and Doc is not happy. We keepers had pushed for leaving the infant with its mother, Pongi, and in this instance, we kept the infant with her far too long. But it is not Pongi's fault. We keepers have messed up, separating her mate, Oscar, from her during and after the birth. In effect, we had changed their normal daily routine and created an environment of uncertainty and instability, and because of that, Pongi was unsettled and not nursing her infant consistently.

We are heading to Doc's house, a twenty-minute drive from the zoo, in the heart of suburban Upper Arlington. This baby will be raised for the next week in his family room until the Birmingham Zoo staff comes to transport the infant back to Pongi's home zoo. We are being abundantly cautious due to our concern about a previous outbreak of salmonella several years prior in the zoo nursery, so best to hedge our bets and temporarily raise the infant at Doc's house.

Doc, in his quiet restrained way, lets me know what he thinks about the baby's condition, and he is not pleased. I hold the infant close to my body, wrapped tightly in towels warmed up in the zoo nursery dryer, trying to bring its temperature up. I'm relieved when we finally arrive, anxious to get the baby settled and truthfully, to get out of Doc's judgmental range. But I know he's right; we screwed this one up big time.

Doc's wife, Mrs. Gardner ("Oh, call me Gertie," she says) is a sweet, smiling, tiny woman. Doc instructs her that she is not allowed down the few stairs from their elevated kitchen into the sunken family room. We routinely keep people away from newborn gorillas, trying to lessen the possibility of transmitting colds or illnesses. I don fresh latex gloves and mask, change into hospital scrubs, then proceed to take the baby's temperature, count his respirations, get a weight, put a diaper on, then give a bottle to get some fluids into him before placing him in the incubator. This is not the first baby I have helped to raise at Doc's home.

Eighteen months earlier in the middle of winter, another infant was born. I was working the evening nursery shift, three to eleven. Darkness comes early at that time of year in central Ohio. It's almost as if there is no defined line between day and night, just different and darkening shades of gray, and as it is January, it is bitterly cold. Early in my shift the snow started.

It is just Mrs. Gardner, me, and the gorilla baby in the house. Mrs. Gardner putters in the living room and then goes into the kitchen to begin prepping supper. She sets the table for three and places a casserole in the oven. Doc arrives home from the zoo, and the three of us share dinner, talking quietly before I return to the family room. Doc checks the baby out, does a thorough once over, and looks at my notes. We discuss any issues or concerns I may have, but we both agree that the infant is thriving, its temperature good, weight gain good, urinating and defecating on a regular basis.

It is now late evening and I am in the Gardner's cozy brown recliner with the infant on my chest, my hand on its back. His breathing is steady—rising and falling—that sweet infant smell comes off his

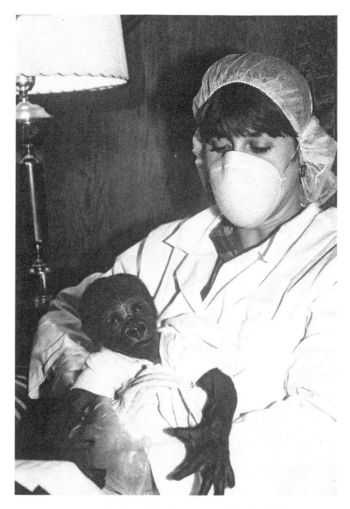

FIGURE 19.1. Beth and infant at Doc Gardner's house

hair coat and rubber-like black skin. Odd tufts of hair comically stand up on his head. I reach up and turn off the nearby lamp allowing me to better see the swirling snowflakes outside. The chair sits in front of the large plate-glass window looking out over the Gardner's back-yard. Within a couple of hours, the snow has become a blizzard and a soft white glow illuminates the darkened room. The whole world outside is enveloped in white, the out-stretched tree limbs cloaked in layers of snow is mesmerizing and a strange winter light emanates from the night sky, casting blue shadows over the yard. I feel as if I am in a cocoon, just the sleeping baby and me. It is one of those rare instants when you realize you would not want to be anywhere else in the world, where time stops and you are simply enfolded in wonderment that you have been placed in this one extraordinary moment, a moment of quiet grace.

20

THE KITCHEN TABLE
AND FRONT BENCH

In most families, the kitchen is a place of comfort, of food, of both emotional and physical sustenance. It is a place to gather, to share stories, to share heartache and happiness, to come together around a communal table, through common cause.

It's no different in the Ape House. It is the heart of the building, the place where we first enter the world of gorillas and where we exit at the end of every day, leaving them to fifteen hours of much-needed privacy. It is a working kitchen as well as our meeting room. It has a stove, refrigerator, countertops, storage space, and bookshelves. We prepare gorilla diets and drinks here and have our weekly meetings. It is a place of creativity and hard work.

Our most valued assets are (1) the long green chalkboard on the wall, used on a daily basis to remind us of changes in behavior or medications or to list our immediate intuitive concerns, and (2) our kitchen table where we read the previous day's Daily Report at the beginning of every day and write a new one at the end of each day. This table is where our ideas come to fruition, everything radiates out from this gathering point. During the warm months, we might use our picnic table decorated with clay pots laden with summer flowers just outside the back kitchen door as our daily meeting space.

After the initial morning routine is complete, after our morning walk-through, after shifting animals for their breakfast, after we have cleaned, and the gorillas' dishes are in a sink of soapy water, we meet back in the kitchen for coffee to share stories, concerns, and thoughts. It's where we plan and then plan some more.

Our husbandry ideas are thrown out on the table to be listened to, dissected, and decided upon—yay or nay. If yay, we carefully work out every possibility. We write up the positives as well as potential downfalls, which we then share with curator Don Winstel for his thoughts and input. If Don is comfortable, we plan the logistics, trying to imagine all possible contingencies, and then we implement.

But all starts with a philosophy. A philosophy serving as our foundation, acting as a course corrector when we stray. If we stay true to our beliefs (some that had yet to be articulated and some that are just starting to surface) and if we adhere to the "Do the right thing for the right reasons" philosophy as our cornerstone, then these seeds of creativity will come to realization. Everything, absolutely everything, flows from that basic premise.

If the kitchen table gave birth to our philosophy and ideas, then the front bench, where we do observations, allows us to further explore those thoughts while taking the lead from the gorillas themselves through the simple act of observing their behaviors. As we work throughout the day, the bench might invite us for a sit when something unusual is happening in any given group, taking a break from hosing and cleaning to enjoy the simple pleasure of observation. Oftentimes that one moment may provide the clue to how we should proceed in the next step of an introduction, or it might point us to a health issue with an animal. We tell new keepers to observe, observe, observe: "Never ever presume anything, the gorillas will tell you through obvious and not so obvious ways what they want, what they need. Never bring your presumption to the fore as that will predictably get someone hurt, either a gorilla or a keeper."

Our observation bench allows us to spend large chunks of time sitting by ourselves or with another keeper, discussing what is happening within the gorilla's lives. Some of the most satisfying moments in my career were spent there. We monitored introductions of females to females, female to males, infants to surrogate mothers and fathers. We watch mothers and babies from the bench and chart nursings and maternal behaviors. It's where we take stock of the intricate social lives of the gorillas in the quiet of the Ape House.

*   *   *

Toni had a long history of pregnancies, and although she was given ample opportunities to rear her young, her babies were almost always pulled. Her experiences with some of her births had not been positive: pain, then being darted, and then waking up to find her baby gone. She had also suffered from full-blown eclampsia, having had convulsions during her first pregnancy. I think she quite rightly associated giving birth with stress and nothing but a whole lot of trouble. No wonder she wanted little to do with these tiny squirming vocal beings when they arrived. We recognized that her past history informed her and that the keeper staff past and present had, to a large degree, created this situation. It was not her fault. So with each birth we tried to give her the time and opportunity to raise her infants, but her behaviors usually deteriorated, oftentimes manifesting itself by rocking and shaking the baby incessantly. She seemed unable to assist the infant in finding the nipple, appeared uncomfortable when it did nurse, and looked exhausted and drawn when the hungry infant began getting fussy again. So she would begin shaking the baby, with the infant protesting even more loudly, which then escalated into more vigorous shaking. It was an endless cycle.

Because of her experiences of being darted with anesthetics when her previous infants were pulled, she was wary. If she saw the veterinarian with the dart gun, she would simply leave the infant and try to get away, although she often looked torn. "Do I leave it, or do I take it with me?" appeared to play across her uncertain face. In the end, she usually left the infant, giving us access to it. We would do a complete physical, rehydrate, and then reintroduce the infant back to Toni as soon as possible. In one particular case, we placed the infant on a bed of wood wool and opened the adjacent door, giving Toni access to the infant. She approached, bent down to look at it, gave it a sniff, and then walked away. She was done, wanting nothing more to do with her baby.

Eventually, Toni would be placed on birth control, but prior to that we had a health crisis with her. Six weeks after giving birth (the infant having been pulled due to incessant shaking, resulting in a hernia), I was giving Toni and another female, Cora, their protein drinks. Toni was the dominant female by virtue of being Sunshine's favorite. Both females approached, and I began to give them their drinks simultaneously, Cora cough-grunted at Toni to get her to back off from the drink. Toni, being who she was, immediately acquiesced and backed away. I was surprised by it, as I had never seen Cora overtly make a play for dominance. I tucked this bit of information away and went on with my work.

Working with gorillas is like putting the pieces of a jigsaw puzzle together. There are hints all around us as to subtle changes within a troop: who is dominant, who wants to be dominant, who isn't feeling well, and who is content. Our job as gorilla keepers is to try to see the snapshots of their everyday life, piece them together, and make the appropriate decisions based on those observations.

Several weeks after this incident, I noticed something odd about Toni. She had quit walking upright and had also stopped facing the wall while standing and fluttering her hands along the sides of her face, which was for her a prelude to regurgitating small bits of food to re-ingest. Although none of these behaviors were in any way normal for most gorillas, for Toni they were a typical part of her somewhat unusual repertoire of behaviors.

I called the veterinarian Dr. Ray Wack, asking him to stop by the Ape House to take a look. Ray and I sat on the front bench watching her, as I explained to Ray about Cora's behavior during the protein drink episode and my concerns about Toni who was now consistently knuckle-walking rather than walking bipedally. Her appetite was good, she was eating OK, she was urinating, but while defecating, it seemed at times that it was painful for her, as she would either bite her hand while cough-grunting or pull stool from her rectum and fling it away. We had seen these same behaviors before with Toni so were treating her with stool softener, but it seemed to be happening much more frequently. When looking at all the behaviors combined, Ray deemed them relevant enough to schedule a physical exam. This is something that no one takes lightly as anesthetizing animals is tricky business. Days of knockdowns (as we referred to them) are always stressful for gorillas and staff alike.

On the day of her exam, we fed and cleaned all the other gorillas and got the building cleared of any obstructions so we could easily carry Toni to the waiting zoo van to transport her to the veterinary hospital. Toni was placed in a sterile chute the day before: no hay, no water, no food, so she knew something was up. Tension was high among the

FIGURE 20.1. Toni, later in life as a surrogate to a youngster, both being given drinks. Photo credit: Beth Armstrong.

other gorillas as well, and when the vets showed up with dart guns, a series of cough-grunts could be heard throughout the building. The chute, being a small confined space, served two purposes. It made it easier for the vets to dart her, and once the effects of the drugs took place, the chute prevented her from falling any great distance. If darted in the living quarters, she could possibly climb to the top, and after the drugs took effect, fall up to fifteen feet, being dead weight hitting the floor (albeit a heavily bedded floor). When possible, the keepers tried giving her the injection with a handheld syringe, which greatly reduced the stress in the building, but this time Toni was onto us.

The vets ran a series of tests, and they found that she had a substantial infection in one of her fallopian tubes and an abscess on her lower back. She was immediately placed on high doses of antibiotics and made a quick recovery. Within weeks she was walking upright again, back to normal—for Toni.

21

HUMANS AND GORILLAS

We humans have a complicated history regarding gorillas, including our fascination with them and our selfish need to want them close. Gorillas mirror us in uncountable ways: our interactions with family members, our investment of time in raising children, supplying guidance and comfort as they grow up and away from us. We mirror one another in our joy and laughter when playing, in our friendships and feuds, in our need for privacy and dignity. They delight us in their similarities while at the same time make us somewhat uncomfortable. We see our humanity in them, but when you are in essence the warden that commonality gives one pause.

If there is one story that speaks to the thorny and poignant relationship we have with gorillas it is that of John Daniel. In 1917 an infant gorilla was captured in the wild and brought back from Africa to a small town in England. The infant was given to a Miss Edna Cunningham who raised the male gorilla as if a family member. Stories vary as to what exactly Miss Cunningham knew when she eventually sold her beloved four-year-old John Daniel to a representative of the Ringling Circus. Some say she was led to believe that John Daniel would live at a private estate somewhere in Florida. John Ringling was looking into purchasing land at that time in Sarasota, Florida, so she may not exactly had been lied to, but perhaps some things may have been strategically omitted in the telling. Did she have any idea what Ringling's intentions were, that John Daniel was to be placed on display in Madison Square Garden? I think not. Anyone who has worked with gorillas, who has seen and felt the intrinsic connection we have with them, would not in all good conscience do that to what is, in essence, a friend.

John Daniel left England in February or March 1921 and was dead within two months.

On April 17, 1921, the *New York Times* blared the following headline: "Girl Crossing Sea to Tend to Circus Ape: Miss Edna Cunningham Due to Arrive from London to Cheer Lonely Animal." The article continues:

> John Ringling has been worried about his pet animal, which is worth many thousands of dollars to the circus, and as a last resort sent a cable dispatch to Miss Cunningham saying: "John Daniel pining and grieving for you. Can you not come at once? Needless to say, we will deem it a privilege to pay all your expenses. Answer at once."
>
> And Miss Cunningham wired right back that she was sailing on the Celtic which is due in New York the first of the week. Miss Cunningham is the niece of Major John Penny of the British Army. When her father brought John Daniel back from the jungle, a tiny black mite of almost human personality, she made a pet and friend of the strange ape. He was never caged and moved at will about her home, even playing with the children who visited there.
>
> Last January he was purchased by John Ringling and brought to this country a few days before the opening of the circus season. Since then John Daniel has steadily grown more and more miserable. He refused to eat, became weak and lethargic and would creep wearily into a corner of his bed and hide under his blanket when curious crowds came to gaze through the bars at him. But sometimes when only a few persons were there John would come out to the front of his cages[,] sit on his haunches and look at them so reproachfully and sadly that it seemed as if the poor beast must be suffering the pangs of loneliness and humiliation. At least that is the way the circus folk, experts in handling animals interpreted John's melancholia.

He was so ill last week that yesterday he was taken to a
room at the top of the Garden to be by himself and be able
when he wishes to look out of the window and see the clouds
and the sky. When they moved his bed up there and took John
along he grabbed his blanket from the bed[,] climbed up in a
closet of the room and spread the blanket out on a shelf. He
wanted to be as far from people as he could. But after the door
had been closed a while and he was left in peace, he brought
his blanket down, spread it on his bed and went to sleep comfortably for the first time in several weeks.

What struck me most when I first read the article was the palpable
sensitivity of the staff writer. There was not a whiff of sensational-
ism. Gorillas have a tendency to do that, to bring out our untapped
empathy. While researching John Daniel's history I found a short
documentary on YouTube with actual footage of John Daniel and
Miss Cunningham getting into a car. Could it have been his ride to
his departure point for the United States? John Daniel was the epit-
ome of a young male gorilla, a little on the gangly side in appearance,
unsure of himself, looking around seemingly aware that something
was up. Gorillas have an uncanny ability to discern a change, a shift
in plans or atmosphere. They are a bit suspicious in relation to us
humans. And they should be.

Gorillas that age are hard driving play machines, goofy in their
exuberance. They are oftentimes born with little or no body hair and
with spidery looking limbs. Then they slide smoothly into a stage of
being just beautiful, everything in proportion—body, face, and hair
coat. They stay beautiful for the first couple of years and then again
do another slide into an adolescent stage not dissimilar to humans,
a gap-toothed look with new permanent teeth way too big for their
faces, hair tufts perpetually sticking up around their heads, always a
tad on the grimy, sweaty, and little-boy stinky side. They are blissfully
unaware and uncaring that they are a bit of a mess. And then they
shoot out the other side into adulthood and are transformed into
these magnificently noble and gorgeous creatures.

Adolescent gorillas have a bluster about them. There is lots of
stamping of feet, stiff-legged run-bys accompanied by the *pok-pok-
pok* of chest beats—the tough guy routine—but it's all for nothing.

It takes only one annoyed adult female to voice her complaint via a cough-grunt, which elicits a response from the silverback male, who slowly lumbers to his feet, standing stiff-legged with lips pulled in in irritation, and the youngsters fold into themselves, all bravado forgotten as they sheepishly and conveniently find another place to be.

I see all of this in the slightly uncertain demeanor of John Daniel getting into that car. I know the end of the story; he does not. And seventy years later, as I watch this footage, I am on the verge of tears because of it.

John Daniel's body became a point of intense scientific fascination, resulting again in a headline for the *New York Times* with this somewhat sad and telling realization: "Appendix, Kidneys and Brain Curiously Human in Type Say Scientists." His body was stuffed and placed in the American Museum of Natural History in New York City. Even in death, he was still a valuable commodity.

22

OSCAR

The Columbus Zoo line began with wild-caught Baron Macombo and his mate, Millie, who became parents to Colo. Bongo was brought to the zoo in 1958, having been "ordered" from the wild to be a companion to Colo. Beginning in the late 1960s, Colo gave birth to their three offspring—Emmy, 1968; Oscar, 1969; and Toni, 1971—a lineage of some of the finest-looking gorillas I will ever meet. I say this objectively; it is simply the case that some gorillas are beautiful, and some are not. Colo and Bongo's children and grandchildren are all without exception good-looking.

Today is a typical midsummer day, no breeze, just still, hot air. Thankfully, one of the docent's had brought watermelons for the gorillas. Dianna gets a call that the watermelons are over at the front entrance, so she drives the golf cart over to pick them up. We place the fruit in the freezer for a quick cooldown.

Oscar is Bongo and Colo's only son; he is a fully grown silver-back with his dad's impressive physique and his mother's gorgeous face and penetrating rust-brown eyes. In his yard, which is a grassy sunken affair set far lower than the viewing public, there is a rather large wading pool no more than eighteen inches deep.

Today we place the whole watermelon in his outdoor exhibit before letting Oscar and his females out. As soon as the door opens, he spies it, this lovely striped green treat, and runs over. Picking it up, he carries it very much like a football player running for the goal line,

FIGURE 22.1. Oscar being sprayed with water by Beth

watermelon tucked in the crook of his bent arm, heading over to his pool, his envious females following in his wake, but he has a plan.

Oscar loves water, to be sprayed in his mouth or on his arms and chest, his arm hair coat dripping with streams of water, but he especially adores taking a dip in his pool. It is not unusual for us to walk or drive by his yard only to glance over to see this stunningly handsome gorilla in absolute silly mode, sitting waist-deep in the pool, happily splashing, mouth wide open in a play face, shaking his head back and forth as he beats his chest using the water as a part of his display. He scoops handfuls of water up and scrubs his hands gleefully together as if he is cleaning them. Sometimes his feet float, his toes peeking above the surface of the water as if they have a mind of their own.

But right now, watermelon safely ensconced in his arms, he wades into the pool and sits squarely in the middle, far from either of his females who gaze longingly at the watermelon from the pool's edges while he proceeds to crack open his prize with his mouth and then tearing with hands. If lucky and patient enough, the females may be able to grab a bright pink piece as it floats by, a piece that he has somehow missed in his revelry.

FIGURE 22.2. Oscar playing in pool

And Oscar was our first tool user. Dianna had a large log installed in the center of their outdoor enclosure; drilling holes in it, she stuffed each with peanut butter, seeds, and raisins. This is a normal part of any gorilla husbandry program today, but in the early 1980s, it was quite cutting edge. Oscar spent hours sitting in front of his seed log, digging away with his branches for treats. Later when he became a father, his son, Colbi, joined him, watching his every move, begging for a taste, and learning from his dad.

23

HOLIDAYS

It is magical working the Ape House on a holiday, especially Thanksgiving and Christmas. On these days, it is the norm for two keepers to arrive at 6 a.m., but today I am working alone. Today's daily routine will be cursory with minimal cleaning and double feeding, and I'll be out the door to my family by eleven.

The walk across the zoo is cold on this Christmas morning. The grounds are pristine and quiet, illuminated in white. The fresh snow sparkles, unmarked by footsteps other than a rabbit's distinctive print or a bird's funny scribble moving off into the distance. The air is crisp and clean, the sun just emerging, and the snow makes its unique scrunching sound beneath my feet. The zoo feels special, as if it is mine.

Each morning the gorillas are separated into individual areas to allow us to hand-feed them their fruits, drinks, and special dietary needs. This also gives them time to relax with their food, with no competition, no looking over their shoulders to see if someone else wants what they have. And they are especially happy today with their double feeding. There are lots of contented rumblings going on. Toni is so excited she can barely contain herself. Her vocalizations border on some form of gorilla elation. She is "talking" excitedly in a singsong, high-pitched voice, simply beside herself with joy when she sees double the usual amount of fruits and veggies.

Today I have cut up the greens super fine, scattering them throughout their enclosures. This will keep the gorillas busy for a good long time. And some are already methodically shifting wads of hay to peer underneath for goodies. It looks a bit like a game of peek-a-boo with food being the incentive. They receive popcorn, dried cereal, and sunflower seeds daily, but today extra portions are given. Our zoo docents not only bring the keeper staff homemade cookies and fudge for the holidays, but they also bring out-of-season fruits for the gorillas. I raid the fridge and make a huge bin of blueberries, raspberries, blackberries, and raisins, adding some peanuts.

I hose down the front and back keeper aisles, do the dishes, and wipe down the kitchen. My reward for being here on Christmas is to sit on the front aisle bench and observe the gorillas. Shafts of light cast their way through the skylights, reminiscent of light coming through stained-glass windows of a cathedral. The streams of sunlight catch the dancing dust motes the gorillas have disturbed as they look for bits of hidden food or gather up bundles of hay to make the perfect nest.

The building is warm and cozy. The youngsters are busy playing. Some are having a game of chase, up and down ladders, into chutes, and in and out of different rooms. The adults are relaxed; several have a cache of food in front of them they have mined from the litter of hay to eat at their leisure while lounging on thick hay beds. Females with infants keep a close eye on their babies when the youngsters come running by in their reckless manner. The infants are fascinated with the older kids and they want to play, but their moms are more circumspect, knowing how rough the play can get.

There is laughter. Every once in a while an adult elicits a friendly belch vocalization, "naa-hummm," reassuring everyone that all is good with the world. I watch closely as the gorillas interact. Who is a buddy to whom, who is irritating another, who is making a power play for greater dominance. All of this is part of the intricate social fabric of their lives.

Near the bottom, close to where I am sitting, Toni lies flat on her back. The top of her head is facing me so she must crane her neck to look back at me. Her left leg is bent, left foot resting flat on the floor, her right foot and ankle crossed over her left knee. She half turns her head to peer at me, and extending her arm toward me, she shakes it a bit while vocalizing to me, and I crack up. She does it again, as if she knows I will get a kick out of it. She looks like some sort of aristocratic princess benignly surveying her inhabitants, waving her hand languidly from her palace balcony, willing to mix with the masses for a brief time.

In another enclosure, Pongi is the undisputed matriarch of silverback Mumbah's cobbled together group of females, juveniles, and babies. In this same group, Sylvia is raising her latest adopted kid. And although Pongi has a youngster of her own, she is still predictably drawn to all infants and juveniles. Pongi, usually so composed and self-contained and not known for overtly interacting with other females, sits directly across from low-ranking Sylvia. Pongi begins to flap her hands up and down, shaking them in exuberant excitement at the proximity of a baby. She is begging to touch the infant, her behavior at once desperate and comical. Sylvia looks on with a deadpan face—as only a gorilla can—and keeps a firm grip on her kid. It's evident to me that this will not be happening, and Pongi much to her credit recognizes that as well and graciously acquiesces.

There are so many exceptional moments, allowing me to witness the obvious and subtle social interactions that others will never have the opportunity to observe. When I leave the Ape House, done for the day, smelling of gorillas, looking forward to seeing my family, it is snowing again, the sun casting deep blue shadows on the mounds of uneven snow and on my earlier footprints.

24

INTRUSIONS
AND MISTAKES

Although the gorillas always seem pleased to see us in the morning, there's no doubt about it: we are simply visitors to their lives. This was brought home to me one night early in my career. Cora, a five-year-old female, had a horrible cold. Whenever one of the gorillas is sick with the sniffles, a cold or the flu, we make a big pot of pureed onion garlic soup loaded with vitamin C, then serve it steaming hot, hoping it will break up their congestion. I am so worried about Cora that I return to the zoo around 11:30 p.m., parking behind the Ape House. There are few night-lights on zoo grounds so I approach the back door walking carefully as I fumble with my keys. Once I'm in the kitchen, I quietly open cupboards, heat up the soup, and take it around to Cora.

All the gorillas have been in a deep sleep, and it is obvious that I have disturbed them. I call to Cora and she approaches looking sleepy and put upon, her nose still runny and gunky. Silverback Mumbah sits up, getting his bearings during this unexpected interruption. Slightly embarrassed at my intrusion, I give her the soup which she drinks before she ambles sleepy-headed back to her night nest. I feel like a trespasser meddling in their lives, like a small-time thief caught in the act. It is a great lesson for me; once again the gorillas have put me in my place.

One day I need to move Bongo to his small outdoor area so I can clean his indoor enclosure. He won't go and sits straight-backed at the chute door to the outside, frequently glancing back at me. He has the fiercest look of annoyance on his face and something else I can't quite figure out. Confusion perhaps? I, in my ignorance, think he is simply being contrary and stubborn, and I am quickly losing patience with him. I pick up the hose like I might spray near him—a technique I am ashamed to say we used sometimes in the early days. His frustration then turns to rage and he is absolutely furious, as he should be.

I finally take a good hard look at where he is sitting. He is right in front of a closed, locked transfer door that I had failed to unlock and open. He is perfectly willing to go out; I have simply not given him access. I remove the lock and stand back, knowing what is coming. Bongo flings the door with all his might. I feel awful and contrite about my self-absorption, my inattention, and my willingness to jump to the conclusion that Bongo was at fault. Not only was it my fault, but I have also insulted him to his core.

I spend the next few days giving him treats and apologizing to get back in his good graces. I learned a lesson about really being in the moment and noticing all the details. I also learned that when a gorilla won't move, pay attention. There is a reason.

An obsession all keepers have is to double- and triple-check locks on enclosure doors. If an animal escapes, it is most likely because of a keeper's mistake. I have been halfway home when, plagued by doubts, I suddenly turn my car around and go back to the zoo to check locks once again. Another vital aspect of being a keeper is to never lose focus or get distracted. If an animal injures a keeper, it too is usually because the keeper was not paying attention or something distracted them at a crucial moment.

I am working in the north building where Oscar and his females are housed. Having just finished cleaning the front display, I am now spreading their food in their exhibit when the phone rings in the back aisle. In order to get to the phone, I have to walk the perimeter keeper aisle. Its solid block walls on either side form a gently curved half circle to the back holding area.

After hanging up the phone, I begin to shift gorillas from the back to the front exhibit. I first try silverback Oscar. He peeks through the open transfer door to the front, glancing back at me a bit nervously, and then refuses to budge. I try again, no go. So I think, "OK, I'll put female Muke up front first." So I shut Oscar's access door and open Muke's transfer door to the front exhibit.

Muke has been at Columbus only for a short time. She is an unusually large female, on loan from the Saint Louis Zoo. Off she goes to the front room, and like any good keeper I walk around the curving keeper aisle to the enclosure door to double-check the locks. There is a blind spot until you actually turn sharply to the right to face the full-length exhibit door. I turn and there is Muke, right in front of me—no door in between. We come within inches of bumping into one another. Simultaneously, we jump back startled, each of us with a surprised intake of breath. Muke backs up as I swing the door shut, slip the top and bottom locks through, snapping them with a click before I slide down the wall to the floor, babbling to Muke, thanking her over and over again. She sits down right next to me, the closed door between us now—and rumbles and rumbles to me as if in understanding.

The door that leads directly to the outside—to the zoo grounds, is only a mere several feet across from the exhibit door, something Oscar could have easily opened with the simple twist of a knob. The possibilities of what could have happened had Oscar gotten out of the building haunted me. He might have been hurt or he may have hurt a visitor, more out of panic than aggression—think of the fate of Harambe in Cincinnati or Jabari, a male gorilla who got out of his exhibit at the Dallas Zoo, and was shot dead. All these scenarios played in my mind, and all due to my lack of attention. This was a powerful lesson I never forgot. The fact that Oscar clearly made the decision not to go to the front exhibit was based on his concern about the door being left wide open. Oscar saved my ass that day.

25

A FAMILY

"Here comes the baby," sing-songs a child as Bridgette ambles out the long chute to the outdoor exhibit on Labor Day weekend. Bongo, her mate, and father of the infant, had come out first, as silverbacks will do, to make sure the area is safe for his family. Their two-week-old son clings firmly with his hands and feet to his mother's chest, his bum safely snug in one of his mother's hands as she negotiates the chute system and climbs down into the outdoor habitat. Fossey is making his historic debut as the first mother-reared infant at our zoo, an event that is long overdue. We named this baby boy Fossey after pioneering gorilla researcher Dian Fossey. Our Fossey, too, would be a pioneer.

Bridgette, on loan from the Henry Doorly Zoo in Omaha, Nebraska, had a history of her infants being pulled shortly after their birth. She was, in essence, a baby-making machine. This was her seventh infant and only the second she was allowed to rear. In this case, we "owned" this infant, and the entire Columbus Zoo staff agreed that Bridgette would be left alone to rear her son with her mate. And at the age of thirty, Bongo was finally given what had always been his right but had been denied him, the opportunity to be a hands-on father. In the quiet of the night, Bongo witnessed the birth of his fourth child, and this time he was a part of the process, a quiet, comforting presence.

FIGURE 25.1. Bridgette with Fossey

The birth took place in the old building with its renovated enclosures. Ironically, these were the very same that Bongo and Colo had
so unhappily shared for decades, but with the recent renovations, the
building had a completely different feel. It was relaxed and peaceful, and the new environment showed in the gorillas' daily lives. They
were behaving as gorillas.

Around 8 p.m. on the evening of August 16, 1986, Bridgette began
building a nest. Bongo sat calmly nearby, occasionally vocalizing to
her in gentle encouragement. This was the first time we had left the
male with the female during the birth. In the past, we had isolated
the male from the female, with the end result being a female in labor
stressed by her protector being removed from her. What we learned
from forward-thinking primate facilities like Apenheul and Howletts
and from our past mistakes was that a hands-off approach was the
best approach, just stick as closely as possible to routine. Equally
important, we made a conscious effort to limit the number of staff
present leading up to and during the actual birth, which we believed
had substantially added to the stress level of the gorillas in the past.
We were determined to provide a peaceful, relaxed environment

FIGURE 25.2. Bongo, Bridgette, and Fossey in the habitat

for Bridgette and Bongo's infant to be born into. The birth went smoothly, and Bridgette was a savvy mother, as she began to care for Fossey immediately. Fossey, in turn, displayed the two essential elements we always look for in a newborn: a strong grip with both hands and feet and insistent, determined rooting behaviors.

Within twenty-four hours it was clear that Bongo desperately wanted to touch his son. He sat by Bridgette, so close they were touching flank to flank. Fossey was cradled sound asleep in Bridgette's lap. Bongo tentatively reached out his massive leathery index finger in an attempt to touch the baby while looking in the opposite direction. He feigned nonchalance so studied and so intentional that it was droll, as if somehow Bridgette wouldn't notice simply because he was pretending not to be there. But she did notice, and consistently brushed his hand aside until she finally looked at him as if to say, "I've got your number, pal. I see what you are doing." Bongo, busted, initially acted as if nothing unusual had happened, but after so many thwarted attempts, he placed his hand in his mouth, gently biting it as if venting his excitement and frustration, looking like a kid caught with his hand in the cookie jar. Bridgette was the boss and Bongo would be allowed to touch the infant only when and if she deemed it acceptable. Twenty-four hours later, she acquiesced. I watched as Bongo, his body absolutely still, hesitantly extended his index finger and gently stroked the top of his son's head and then softy vocalized to himself while sniffing his son's scent from his fingers.

As time went on, Bridgette allowed Bongo more frequent contact with their son. Sometimes Fossey wandered over to his dad to play with his toes and belly. Bongo sat ramrod straight, unmoving, staring straight ahead never once glancing at his son while he gently reached down to touch him. All the while Bridgette was in the background ever vigilant, watching Bongo's every move.

26

BUILDING A GORILLA TROOP

Discussions began in earnest in 1986 about creating an age-diversified troop. We had several adult male gorillas, and this allowed our females to choose which male they felt most comfortable partnering with. In essence, we had options. We also had numerous females that had arrived from other facilities in the last few years: Bridgette, from Henry Doorly Zoo in Omaha; Lulu, from Central Park Zoo in New York; Pongi from Birmingham Zoo in Alabama; and Sylvia, from Baltimore Zoo via the National Zoo in DC. Our male gorillas included twelve-year-old Sunshine from the San Francisco Zoo, Mumbah from Howletts in the UK, and Bongo and his son, seventeen-year-old Oscar. All were unique in their own way, but many shared a common thread of having been captured from the wild as an infant. Those memories, those experiences surely were still there, informing each of them.

By the time I came back to the Ape House in June 1986, the idea of incorporating the twins, Mac and Mosuba, into Mumbah's troop, which included their grandmother, Colo, and their half sibling, seven-year-old Cora, was already being enacted.

The issue with the boys was that our agreement with Omaha's Henry Doorly Zoo still held. We had the twins five to six months out of the year, and then they were transported back to Omaha. We felt that we could offer them a more stable social environment but negotiations with Omaha continued with no real agreement in sight.

So we worked around the twins' travel schedule over the next year until they were fully and finally integrated into Mumbah's troop by July 1987. We recognized that something would need to be done and a decision made because these constant disruptions of traveling back and forth were not only stressful for the boys but on Mumbah's troop members as well. The dynamics changed every time the boys were taken away. We were asking a lot of the troop members to constantly adjust to these changes.

While sitting at the kitchen table one morning, I threw out the idea of creating an age-diversified group. I had been kicking around this idea for a while during late-night brainstorming sessions at Charlene's house, and together we had looked at it from all different angles. My experience observing snow monkeys at Columbus Zoo and Apenheul's gorilla troop validated my belief that infants and juveniles were the glue that held any primate group together. The presence of an infant gave the adults their jobs back—overseeing, protecting, and guiding a vulnerable baby. The infant would benefit from being raised from a young age with conspecifics learning the dos and don'ts of gorilla social life. As importantly, rambunctious juveniles would be exposed to and learn gentleness in regard to a young infant. The adults would intervene when play became too rough, thereby teaching the juveniles restraint. This perception was not exclusive to me; we all saw the possibilities and realized that this innovation might be used as a model for other zoos, not just for our gorillas at Columbus.

Our goal was to create a troop that closely mirrored those of the wild: a silverback, several adult females, and numerous juveniles and infants. How we got there was open for debate, but rather than waiting for the next generation of mother-reared infants (which were still few and far between—at least in the United States) to raise their own young, we decided to circumvent the process, speeding it up and artificially creating an entire troop. We would expose young nursery-reared gorillas to proper gorilla etiquette at a very young age. Thus our Gorilla Surrogate Program was born.

My observations at Apenheul showed me a gorilla troop with vitality, constantly in motion. Even in the afternoon, while many of the adults tried to nap, the juveniles and infants were busy playing and roughhousing. Our intent of creating an age-diversified troop was to break the cycle of having to hand rear another generation of infants, which we did all because their mothers, nursery reared themselves, had no exposure to young gorillas. In essence, we wanted to shatter the status quo of socially isolated gorillas so the next generation of nursery-raised infants and juveniles introduced to and raised in a troop setting would theoretically go on to raise their own offspring, thereby breaking the decades-long cycle. We believed that gorillas were incredibly adaptive, not prone to drama. We believed if given an opportunity to live in an environment that felt safe, private, and comfortable, they would invariably behave as gorillas, regardless of their backgrounds and social history.

When we began to discuss the surrogate program, we were already committed to mother rearing and were actively working toward that goal with the upcoming birth of Bridgette and Bongo's baby, Fossey. In 1984 head keeper Pete Halliday had arrived from Howletts in the UK, bringing with him Mumbah, who was then a nineteen-year-old silverback. Mumbah was small for an adult male and had an odd elongated mouth that often hung open. Mumbah's eyes always had a weariness to them. His reactions were mild in relation to his surroundings, as if he wasn't always fully engaged. Strangely, he would become hypervigilant at the oddest of things, such as a large piece of unfamiliar machinery parked near the Ape House. Such things could throw him for an absolute loop, resulting in alarm barks, and then he might not come back in the building at the end of the day. But we watched his behavior around other gorillas and felt that he would make a good candidate as a surrogate father to very young infants. Make no mistake about it, without Mumbah we would not have had a surrogate program. Every group member had a vital role to play, but he was the foundation.

So many essential factors came into play with enacting this concept and what was logistically needed to allow for success. We could do introductions in the privacy of a restricted building, no public viewing. Because our enclosures were three-dimensional, they allowed for more options and movement for the gorillas. To further enhance their choices we had filled blank walls with foot- and handholds and had additional climbing structures. We were given free rein to figure out the logistics of how to start building this troop. Mumbah, adult females Colo and Cora, and the juvenile twins would form the foundational core of the troop; now it was time to start building with the inclusion of infants. We were very mindful of how we wanted to approach this process; we carefully laid out preliminary plans but added a generous dose of openness. We were willing to change and adapt as the gorillas let us know what worked and what didn't.

In early 1987 while Bridgette and Bongo were raising their five-month-old son, Fossey, another infant was born, JJ. When Toni did not raise him, we began working out a step-by-step plan of how we could integrate JJ into Mumbah's troop. In preparation, JJ was brought to the Ape House daily to acclimate him to the sights, sounds, and smells of gorillas.

Starting years before, the head keeper of the Children's Zoo and nursery, Dusty Lombardi, and I had had numerous discussions about incorporating gorilla behaviors into daily nursery protocol and husbandry. Infants were to be treated as gorillas, not humans, and as such JJ was carried on the nursery keeper's chest. When older, JJ was encouraged to climb onto and ride on the keeper's back. Fruits and vegetables were introduced into his diet early on and the keepers ate with him on the floor, mimicking feeding vocalizations. JJ was encouraged by the friendly, "mmm-wahhh" vocalization and was disciplined using cough-grunts. Every day, the keepers engaged in play sessions tickling him in favorite places—neck, armpits, and on either side of his groin.

In a 1996 article in the *Gorilla Gazette,* I wrote:

The Columbus Zoo husbandry program dates its many successes back to the birth of the first gorilla born in captivity in 1956 and on to the birth of surviving twins in 1983. But these accomplishments can be attributed as much to luck as to any conscious husbandry program. In the mid 1980s however, the Columbus Zoo staff came to the conclusion that reproduction alone was not far reaching enough in its goal but rather the early socialization of all infants was in the best interest of the next generation of gorillas. Taking our cue from facilities like

Howletts in England and Apenheul in the Netherlands where mother rearing was the norm rather than the exception, we decided that age diversity was the key to enhancing the lives of captive gorillas. Our philosophy derived from a belief that gorillas had in large measure failed to rear their own offspring not because of their inability, but as a result of our collective lack of understanding of gorilla behavior and husbandry.

Finally in March 1988, we began the introduction process of fourteen-month-old JJ to Mumbah's troop. Leading up to this crucial step, we had tried a couple of different approaches. We allowed the twins access to JJ, with two keepers present, one nursery keeper holding JJ and a gorilla keeper sitting nearby to intervene if necessary. As some of us predicted, it was an absolute mess. The twins gleefully spied a new plaything, were too rough and tumble with JJ, insistently and stubbornly trying to peel him off the nursery keeper. JJ for his part hunkered down with his head tucked in, his body tightly curled into his keeper's body, making himself as small as possible. The boys loved it, but JJ not so much. The good thing that came out of that brief interlude was that it only served to confirm what we already believed: the presence of adult gorillas was an absolute necessity. They needed to be in the mix from the get-go, disciplining rowdy youngsters in relation to infants. Something else became very evident; the twins were spoiled rotten, fully indulged by their troop members, having never been disciplined by Mumbah, Colo, or Cora.

Colo, JJ's grandmother, chose to be his adoptive mom. We conducted extensive observations, which clearly indicated that she spent considerable time at the mesh watching JJ on his daily visits, vocalizing to him, passing items to him, such as hay, food, and small pieces of cloth. All of these behaviors indicated a keen interest on her part. This is not exclusive to females. Keepers have seen males also express an interest in infants. In fact, a few years after Colo and JJ's successful introduction, a silverback male named Fred at the Saint Louis Zoo became the surrogate parent for three male youngsters. In a 1990 article in the *Gorilla Gazette,* Ingrid Porton, curator at the Saint Louis Zoo, wrote, "Although obviously interested in the infants, Fred did not force them to interact with him. He sat or lay down quietly and allowed the infants to come to him. He remained motionless

FIGURE 26.1. Colo with JJ on her back

when the infants tentatively touched him and quickly moved away. Occasionally he slowly reached out a hand to gently touch an infant."

After months of careful observations and much discussion among keepers and the curator, we finally decided to open the doors between Colo and JJ. JJ immediately walked over to Colo and tried to climb onto his grandmother's back. Colo was not too keen on that and actually seemed to flinch; she kept removing him by sliding her hand down under him, gently but insistently peeling him off until JJ got the message. But as their time together progressed that day; she began to get quite rough with JJ, tossing him about on a couple of occasions, so we ended the introduction before the roughness could escalate. The next day fared better, as they spent the day sitting close to one another, both relaxed. On the third day we decided to leave them together overnight. Day 4, all looked good when we came in that morning and we breathed a collective sigh of relief.

Colo and JJ were then given time to form a bond and establish a routine. They slept and ate together, transferred together, working out their own daily dance of sorts. And every day their bond strengthened as they better learned each other's patterns and personalities.

Next step, the twins and Cora were included—and all went well. But one day shortly after the initial inclusion of the twins, the inevitable happened. One of the twins grabbed JJ, dragging him across the floor. JJ was quiet initially, trying to roll up into a ball, but when Mac let go, Mosuba swooped in as if on cue grabbing JJ and dragging him back in the opposite direction. JJ began whimpering, and then hooting as the boys continued on and on, back and forth, back and forth. The twin boys were having a great time, escalating their roughness, oblivious to any ramifications. They were, after all, the much-loved and indulged youngsters in the troop.

Colo ignored them initially, but once JJ started to sound truly distressed, she reluctantly got up, then gaining speed like a locomotive, Colo moved quickly to the offending twin who was blissfully oblivious to her, still happily dragging JJ around the floor. Suffice it to say he didn't see it coming. Colo began cough-grunting at the twin before she even reached him, but he ignored it. It wasn't until Colo grabbed him, threw him down on his back, hovered over him, and then proceeded to mouth him, did it seem to dawn on him, "Holy crap, Granny's pissed, I'm in deep trouble." Bear in mind, Colo had very few teeth at this stage of her life, so whatever she was doing could not have been physically hurting him, but it scared the piss right out of him, literally.

She continued to hold him down, cough-grunting, not letting up as he released an arc of urine, while simultaneously shrieking at the top of his lungs, frantically trying to scramble away from her. Once she released him, he scurried crab-like away to the adjacent enclosure followed by his twin brother, both hooting pitifully and clinging to one another. His dignity was somewhat bruised, but the twins had both learned a valuable lesson. This did not mean, by the way, that they did not try and tag-team JJ in the future—they did, but usually outside where there was a lot more room and Granny could not get to them as easily or as quickly. They would furtively look around to see where Colo was before embarking on their forays of torment.

But the twins also had another side to their personalities. In late fall of that same year, we had installed clear plastic flaps on the overhead chute exit doors in preparation for winter—an idea we appropriated from Apenheul. This allowed us to give the gorilla's access to the outside habitat on cold days, while at the same time the heavy

FIGURE 26.2. JJ riding on a twin's back

flaps retained heat in the building. JJ had never used the flaps before. As I was shifting the group back into the building on that chilly
gray day, JJ was the last to go in. For some inexplicable reason, Colo
had left him behind, something she had never done before. JJ sat in
the outside transfer chute softly hooting to himself. Poor guy couldn't
figure out how to get back in.

On the way out, it is relatively simple to deal with the flaps. The
gorillas just push through the flap using their body weight and forward momentum. But on the way back inside it actually requires
some effort and the gorillas have to lift the flap up and duck under
it. I was standing under the chute quietly reassuring JJ while formulating a plan as to how to get him into the building when one of the
twins came back out. I watched as he gently put his arm around JJ's
shoulder and walked him down to the flap, lifted it, and JJ scooted
on in. These gorillas, both young and old, constantly amaze and
charm me with their solicitous behaviors toward more vulnerable
troop members.

After each subsequent member was introduced to Colo and JJ, we
gave all of them time to form a cohesive bond before adding another

troop member. As this was new territory for us, we hoped that all the troop members would come to JJ's aid should there be a problem when we introduced the silverback Mumbah. As far as we knew, no one had introduced an unrelated adult male to an infant in captivity, and although Mumbah was about as easygoing as they come, infanticide was a very real possibility. As a staff, we agreed that timing was everything, that it would be just as problematic to go to the next phase of an introduction too soon as it would be to wait too long. We couldn't explain it, but we somehow knew when it was the right time to introduce the next member.

The great thing about this program was that we had four voices gathered at the kitchen table to go over the pros and cons, discuss timing, what to do next in the midst of intros. Each of the keepers, Adele, Dianna, Charlene, and I brought their own insights, experiences, concerns as well as their gut feelings, which were an essential part. We discussed the what-if scenarios and had a contingency plan in place should something go wrong. Mumbah's inclusion in the group was the big step, and one particular day we all just knew it was time.

During any introduction, we always have one keeper in the front and another in the back aisle with easy access to a hose should a fight erupt, and we have to quickly spray water into the enclosure to break it up if the intro got out of hand and became too rough. The other two keepers sat on the observation bench taking notes.

Our zoo photographer, Nancy Staley, was there to film this crucial next phase of the introduction. As Nancy was at the Ape House every day, I think the gorillas just looked at her as our fifth keeper; they were so used to her that she had no discernible effect on their behaviors. We would never have been able to get such fantastic footage of introductions, nursings and behavioral changes without Nancy's presence.

When doing intros, the four of us were like a well-oiled machine. We got the morning cleaning and feeding done quickly and efficiently. This enabled us to start the introductions early in the day, which in turn allowed us to observe them for a good four to five hours. This was critical, as our decision at the end of the day about whether to separate the gorillas from one another for the night would be based on these hours of observations. Ideally, if all looked positive and calm, we would simply leave them all together overnight. The next morning when we came in and all was well—that was it, end of introduction.

We prepped the introduction areas with extra hay, loads of treats, and enrichment items—cereal boxes or paper towel tubes filled with goodies like raisins, cereal and popcorn, fruit yogurts and honey drizzled through the yellow pages of the telephone book, and honey and yogurt drizzled down the sides of the mesh caging. It served two purposes: (1) it kept the gorillas occupied, and (2) introductions were associated with a positive experience for all.

Once we were all in position, Dianna called for the doors to be opened, and the keepers at the back and front opened the doors simultaneously. We could never have envisioned in a million years what we were about to witness. Thankfully, Nancy was there to film it all. This footage would be used in future lectures, presentations, and a documentary to show just how utterly adaptable gorillas are and how introducing infants was a doable husbandry tool.

Mumbah slowly walked into where Colo, JJ, Cora, and the twins were, his face and body relaxed. On the other hand, we the keepers were a bundle of tense nerves; things could turn quickly if Mumbah became aggressive. We were in completely new and uncharted territory. As Mumbah slowly approached sixteen-month-old JJ, who was on the floor, the rest of the group members quietly surrounded JJ in a protective force field, all carefully watching Mumbah's body language. Cora hightailed it over from the back part of the enclosure. Colo, of course, was behind JJ, backing him. Most astounding were the twins, who came up on either side of JJ as he slowly moved toward Mumbah. And just as JJ looked up into Mumbah's face, Mumbah simultaneously leaned down to peer into JJ's. It was truly breathtaking. Then one of the twins did a displacement behavior, very subtly brushing up against Mumbah trying to distract him. It's a social mechanism used in tense situations to thwart a potential altercation. And then Mumbah quietly sat down directly across from JJ. You could almost see the group relax in unison, while we, the keepers, let out a sigh of relief. That was it, JJ was accepted, and the troop was complete. The entire process from start to finish, from the day we opened the door for Colo and JJ to the inclusion of Mumbah was less than two months. Now we were on a roll. We knew Mumbah would accept infants, so we were already thinking ahead, planning our next

FIGURE 26.3. JJ and Mumbah sitting next to one another in the habitat

addition. As a staff, we were more than ever determined to create an age-diversified troop using adoptive mothers, father, and siblings.

From March 1987 to early 1996, when I left the gorilla department, we had created a fifteen-member troop consisting of:

1 silverback—Mumbah
3 adult females—Colo, Sylvia, Bathsheba (Cora was eventually moved to Sunshine's troop)
2 adult pregnant females—Lulu, Pongi
3 juveniles—Mac, Fossey, Colbi
2 infants born into the troop—Kebi, Cassie
4 adopted nursery-reared infants—JJ, Jumoke, Nkosi, Nia

Our program began getting wider attention in the early 1990s, and other zoos began sending their infants to us in the hope they could be integrated into Mumbah's troop. But over the years, I became concerned that perhaps we were not addressing the very real issues at other institutions that were contributing to this influx of infants: mainly that these zoos might not be doing everything within their ability to retro-fit enclosures, to change their husbandry program, or to shut down public viewing after a birth.

I felt compelled to write an article for *Gorilla Gazette* addressing those concerns. Because each successive introduction we did at Columbus resulted in our ability to integrate infants at a younger and younger age, perhaps we were unwittingly offering an alternative to other zoos to mother rearing, but that was never our intent. Our successes seemed to have an unintended consequence. We started the surrogate program to put an end to mothers not rearing their offspring by the simple fact that—if surrogacy was successful, and it was in Columbus—these infants would become proper gorillas at a young age, schooled in the ways of gorilla etiquette by mature gorillas, and would go on to raise their own young in the not too distant future.

Our intention was, first, to improve the lives of the gorillas directly under our care at Columbus by creating the surrogate troop and, second, to share this husbandry example with the zoo world. We hoped that the overall philosophy of promoting mother rearing, combined with surrogacy as a bridge when needed, might be adopted by other institutions. In the process, Columbus created a trend, a model that could possibly shift the culture of zoos toward all infants being raised by gorillas.

27

CREATIVE SPARKS

Painter Henri Matisse said, "Creativity takes courage." What is it that prompts creative thought? Is it something in the air, in the timing, the setting, the people that surround you, the apparent need for change, or all of the above? All are individual components reacting to and against one another, eventually igniting an idea. I think that creativity feeds on itself, that one idea leads to another and another and then another. It is as if alchemy is at work, that all of these elements play and push against one another, sparking new ways of looking at old problems. One element that is essential to the process is courage, courage to be willing to look at things differently, to come at a problem from a dissimilar angle, and to push back on the established norms and trends in order to affect change.

Going through daily records from the 1980s, I find there were two periods of incredibly fruitful change within the Ape House. The construction of the Gorilla Habitat and the completion of renovations of the old building in 1984 resulting in the capacity for us to house more gorillas. What began as a trickle that year turned into a welcome deluge of gorillas arriving over the ensuing years. I also noticed that 1987 and 1988 were years of play and fluidity in the group structures, both indicators of the health and complexities that were taking place within our four gorilla troops.

So many events acted in our favor. The Gorilla Species Survival Plan (SSP), which now oversees where and when gorillas are sent from one zoo to another, was created in the early 1980s but was not fully in effect until 1988. Until that time, it was a free-for-all for us in terms of receiving gorillas from other zoos. Because so many gorillas had yet to reproduce, genetics were hitherto not a factor at that point, so we were not hampered by outside recommendations. We were once again fortunate victims of circumstance in our timing. We had a blank canvas on which to start creating gorilla troops. Many factors go into why group dynamics work or don't work, but one of the most important facets was allowing each of our newly arrived female gorillas to make her own choice as to where she wanted to be placed and with whom she wished to live.

New arrivals went through a quarantine period before being brought down to the Ape House. Once there, we gave them time to get used to their living space and surroundings, to get settled in and become familiar with the daily feeding and shifting routines. Then we watched carefully to see how they reacted toward and interacted with other gorillas. If a female was interested in a specific male, she might elicit a breeding solicitation call to him or vice versa. If the other returned the interest, we would then place them next to each other and watch their interactions to see if they shared food or play items, touched and tickled through the mesh, or vocalized to one another. All of these observations went into our decision-making as to where a new gorilla would be placed. The key point was that everything we did was responsive. If a female decided later that she didn't want to be in her troop any longer, we listened to her and gave her what she wanted by removing her and placing her with another male. It was an ever-changing organic process.

Our notes from 1988 show time and again the flexibility of our daily decisions. An already established female might one day be placed with a new female for companionship during the day—but the next day that same female might be placed with a male from a different troop in order to breed, because the two had exhibited interest in one another, and then be returned to her home troop at the end of the day. Our notes show that sixteen-month-old Fossey and eleven-month-old hand-reared JJ were given access to one another on a daily basis for play sessions while Bongo watched from an adjacent area. Although Bongo was initially worried about being separated from his son, once he came to trust that it was only temporary, he seemed to relax, but it wasn't beyond him to elicit an "oo-oo-oo" cough-grunt from his vantage point if he thought the play was getting too rough.

While rereading the 1987 records, I noted endless daily notations on play sessions happening all over the Ape House. Play is indicative of the health of any troop, reflects a relaxed environment, and indicates age diversity because kids play no matter what kind of primate they may be. Because we had infants and varying ages of juveniles, we were starting to see a true reflection of what one might see in the wild—constant activity, play-chase, adults intervening if things got a little out of hand, and even adults much more active, being more playful and silly themselves.

Notes from Daily Records

February 1987
—lots of play in all groups.
—Bongo upside down on rope & swinging.

March 1987
—very active today.
—gorillas playing, vocalizing.
—lots of play, interacting in gorilla groups.

April 1987
—Toni playing with everyone!
—Toni & Lulu playing (2 adult females); Bongo and Bridgette (mates & parents to Fossey) rolling & tickling each other.
—Bongo playing with Fossey (his 8-month old son), Bridgette pulled Fossey away & initiated play with Bongo.

May 1987
—Cora & Fossey playing through mesh a.m.
—Bridgette sitting next to Fossey while this interaction occurring; Bridgette chasing Fossey—both laughing; Fossey playing "tug of war" with Bongo using burlap.
—Toni & Sunshine hugging & playing in exhibit A. Shine reached out to Toni to initiate.

The year 1987 continued to be a period of incredible growth and creativity within the department as we began to explore projects unrelated to the daily care of gorillas but broader in scope and with a more far-reaching effect. Our photographer, Nancy, and Julie Estadt of our marketing department came up with the idea of producing a gorilla calendar highlighting our gorillas and their individual histories to be sold as a fund-raiser. I brought up the idea of creating and publishing a newsletter called *Gorilla Gazette* in February. We got to work and by July, we had our first issue wrapped and in the mail. Another July milestone: the twins were fully integrated back into Mumbah's troop. In early January, JJ (Jungle Jack), son of Toni and Sunshine, was born but was pulled and was being raised in the gorilla nursery. His birth and decision to raise him in the nursery prompted subsequent discussions about creating the Gorilla Surrogate Program, which took root and began to grow.

I wrote my first paper in 1987, which was about Fossey's birth and our underlying commitment to mother/father-reared babies. I found that I loved the creative process of coming up with an idea, thinking it through, doing a brief outline, and then writing the paper. Once done with the preliminary draft, Charlene and I would then fine-tune it at her house late into the evenings. We carefully selected slides that would best illustrate and convey the story we were trying to tell. Then I timed Charlene as she read it over and over again, until it was polished and ready for presentation.

This, our very first paper, was to be presented at The Ohio State University at a gathering of anthropologists in April 1987. Later that same year, Charlene and Nancy would travel to Seattle, Washington, to give the talk again at the national zoo conference. But today I took time off in order to attend the presentation at OSU and because it was a Saturday, the last day of my workweek, it already had a distinctly holiday feel to it, as if it were a Holy Day like those I remember as a Catholic school child—a gift.

Charlene's presentation went off without a hitch and was incredibly well received, if the extended length of the Q and A session was any indicator. I think its positive reception was because we were presenting practical knowledge based on daily observations; there was nothing theoretical or academic about it. Only real-life, real-time experiences were presented. Working on that paper solidified my love of writing, of organizing my thoughts and creating stories. It also confirmed to both of us that it is the stories you tell—not necessarily the data or the graphs you present—but the touching and relatable stories you share that can change people's perceptions.

On the drive home, while basking in the success of the paper, another gift was dropped on me, the magic of an unexpected spring blizzard. Sitting at a stoplight, I watched as the traffic lights were blown horizontally and felt my car shudder in the relentless winds. I skidded my way on the slippery streets, relieved to make it home safely. My red brick apartment complex looked like some sort of otherworldly fairyland, dipped in a coating of glittering white snow. After hanging up my coat and changing into sweats, I put on the kettle and made a cup of tea. Sitting in my cozy kitchen I looked out my window in deep appreciation and wonderment at the beauty that surrounded me as I reflected on such an enchanting gift of a day.

Perhaps no other place was more important to helping shape the gorilla husbandry program at Columbus than Charlene's kitchen table. I have already mentioned the table in the Ape House kitchen and the picnic table out back. But after hours, and after my second job, I often headed over to Charlene's house. Her husband, Bobby, would start the evening by making us a fresh pot of coffee. Over endless cups we planned and dreamed, finding ways to surmount any obstacles we might find in our path. As in any workplace, all was not smooth in the Ape House, and sometimes we had to figure out ways to get past any possible reluctance Dianna may have had to a new idea. And truthfully, in defense of Dianna's reluctance, we were coming up with some pretty radical ideas for the times. Looking back on it, I realize now that having to jump hurdles was not such a bad thing. The resistance made us more imaginative. Sometimes the solution was as simple as, "You bring it up because she'll listen to you" or "You bring it up first, then I'll write something up and present it tomorrow." We learned that how you present an idea may very well be just as important as the substance of the idea, because ultimately if the idea was not allowed to take root and grow then our ideas were futile so we maneuvered our way through.

You're fortunate if you have a coworker who is as passionate about change, about making a difference, about being part of something bigger than yourself. I was blessed to have Charlene as an equal partner. Our individual strengths supported one another—where one had a weakness, the other filled in the blank space until we could catch up, balancing and supporting one another. It was truly a team effort. All of this, this willingness to speak out, coupled with a husbandry program that could only be improved upon, added to our mutual respect and love for these unique creatures within our care, and an imaginative spirit within each of us made for a team that was unstoppable. Although frustrated at times, we always found a way, the work got done, and the lives of these animals improved and became more complex and fulfilling.

While changes were abounding in the Ape House in terms of improvement in husbandry, the staff realized that while what we were doing at Columbus was directly benefitting the gorillas within our care, that it was innovative and groundbreaking, these innovations could actually have a much greater effect on the captive population in North American zoos. In essence, our intent was not only to give back to the Columbus Zoo gorillas but to share our knowledge and experiences in order to affect a change in philosophy and husbandry techniques at other institutions. So we began writing additional articles for *Gorilla Gazette,* presented more papers at a variety of conferences, and solicited specific articles for *GG* from institutions that we thought were more forward-thinking and could possibly spark and nurture this change.

If 1987 was a year of play, then 1988 was year of continual and monumental changes. March, in particular, was busy. We began the introduction process of JJ to Colo, thus laying the practical foundation for our surrogate program. Also in March, Lulu gave birth to her daughter Binti. Around the same period of time, I flew out west to bring in an additional female named Bathsheba from another zoo. By 1988 Adela Absi was hired into the department on a full-time basis and with her our team was complete. Each of us brought our own strengths, weaknesses, and different perspectives to the table, and it worked. If one of us felt discouraged or burnt out for whatever reason, another picked up the slack until we were ready to dive back in. The four of us were a formidable force. We may not have always agreed on everything, but we figured out a way to get past barriers.

In spring of 1988, I began my four-year journey of working toward a bachelor's degree in anthropology, working all day at the zoo and then attending classes four nights a week. So to say it was a busy time is an understatement.

1988 Daily Records

18 April 1988
—Colo and JJ introduced to twins, Mac & Mosuba.

20 April 1988
—Binti (infant) rode on Lulu's back; Intro of twins to Colo/JJ
    continues.

1 May 1988
—Lulu's Binti nursed throughout day.
—Colbi (8 months old) walked 15 ft. away from Pongi.
—Bathsheba (new female) spent 2 hours in exhibit next to
    Bongo and Fos—she sat at mesh, she and Fossey touched
    while Bongo sat 5 ft. away—all relaxed.

2 May 1988
—Oscar/Pongi/Colbi brought to habitat.

3 May 1988
—Infant's (Binti) color good, alert, active. Stool from infant
    found in night exhibit.
—Bathsheba next to Bongo/Fossey for 3 hours. Fossey and
    Bathsheba playing through mesh. Bongo watching from
    4 feet away.
—Pongi quietly approaches Sylvia in the yard—Sylvia consis-
    tently moving away.
—Mumbah spent some time at the mesh watching Cora/
    Colo/JJ.

4 May 1988

—Lulu's infant's (Binti) stool seen in night exhibit (gold and pasty glue-like). Infant nursed throughout the day. Infant alert, color good. Infant riding on Lulu's back.

—Mumbah playing with twins with Cora standing next to Mumbah.

—Colo/Cora/Twins—filmed together for one and half hours. Twins playfully aggressive, many times JJ became frustrated at this behavior. Cora not involved in either play or discipline.

—Bathsheba spent two hours next to Bongo & Fossey. Fos/ Bath touching through mesh.

11 May 1988

—Mumbah & twins playing, Mumbah being chased by twins, rolling, laughing.

—Colo carrying JJ ventral/ventral.

—Cora & Bathsheba together in p.m.

28

LEARNING TO BE
A GORILLA

Fossey is tottering away from his mom today. He sees his dad in the distance and is inevitably and inextricably drawn to him. Bridgette eyeballs the situation and then seems to sigh and heaves her substantial bulk up to go gather her infant back into her arms to keep him tethered close. This will eventually change, and Bridge will allow Fossey more frequent time with his father, but not just yet.

Months later the infant phase is over; Fossey's time glued to his mother's chest is officially done, and the toddler phase has begun. She has started shifting him to her back when moving around their daily lives. Back transport can be quite entertaining to watch. In the beginning, infants will lie with their chest flat on their mom's back, a ventral/dorsal (infant's chest to mom's back) position holding on for dear life with hands and feet. Once the kid gains more confidence, the infant may sit upright on the mother's back looking like a jockey atop a horse, rhythmically moving to the mom's gait. And in what I can only describe as cloud gazing, an infant might be seen to lie flat on his back, idly lollygagging, gazing up at the sky. When it is time to shift locations, to transfer in or out, youngsters get it in gear and rush over to their mother to climb on board, anxious not to be left behind.

Bridge has devised different strategies to address this new toddler phase of motherhood. Naptime is a bit problematic, but she has discovered that if she grasps Fossey by an ankle, she can take a quick snooze while he moves only as far as her arm can reach, in a half circle. When

FIGURE 28.1. Fossey in tub as Bongo looks on

Fossey was really tiny, she used to load up one of the high-sided black rubber tubs scattered throughout the exhibits with soft alfalfa hay, then place Fossey down in the tub while she dozed nearby.

Over the last couple of months, we have noticed something else. Bridgette, who simply adores food and has the body to show for it, drops bits and pieces of celery, lettuce, and apples on her chest where Fossey clings. Fossey takes to sniffing and exploring the pieces before tentatively mouthing them. She is doing this intentionally; she is

FIGURE 28.2. Kebi teething on wire mesh

schooling him. Recently, Fos grabbed a long stick of celery and proceeded to bite all along its length, experimenting. He didn't eat any, but he seemed to be testing and tasting, perhaps enjoying the distinct crunching sounds he was producing. Or he may have just been teething as any number of objects are used during the teething phase, including the wire mesh. Fossey's development is on target, and he is slowly but surely adding solids to his diet.

In addition to the usual developmental stages of growth for infants—crawling, erupting teeth, eating solid foods, spending more time away from mom—there are some social norms that are learned and must be adhered to. Being polite is one. The basics are: watch your vocalizations and watch your body language.

I was outside giving a midafternoon snack one day when I watched the twins begging from their grandmother, Colo. There is nothing quite as diverting as a juvenile gorilla, usually so full of unbridled energy, who wants a treat from an adult but sits in restrained longing gazing at the adult. The twins sat as close to Colo as possible.

FIGURE 28.3. Juveniles begging from Colo

Any closer and they would have been sitting in her lap. Colo wears a practiced patient look that basically says, "I know you are here, but I'm choosing to ignore you and will continue to do so until every last crumb is gone." She looks past them as if they don't exist, as if she doesn't notice that they are leaning in just inches from her mouth. Instead, she concentrates on her food. The twins are close enough to get a good whiff of whatever it is she is eating and they seem to close their eyes at times, just breathing in the fragrance of said food morsel. But they have good manners so do not reach out or grab. They have been taught well. Finally, when Colo is done eating, the boys take off—looking momentarily disappointed, but they are resilient and quickly get back to their busy life.

29

KEEPER ETIQUETTE

I always make sure to whistle when I walk the perimeter keeper aisle, especially down the back aisle. It is a courtesy to let the gorillas know I'm coming so I don't spook them by walking up on them unexpectedly. Think of being lost in a good book, or bending over to put a load of laundry in, or standing at the sink doing the dishes when someone walks up behind you. Because you are engrossed in what you are doing, you nearly jump out of your skin, even sometimes lashing out angrily in your fright and adrenalin rush. It's a reaction that is perfectly understandable, and it is no different for them.

It's our job as keepers to always know where we are in relation to them in recognition that it is their home we are walking through. I tell new keepers or visitors to the building, "Think about this. You are in your house, and all of a sudden without warning, a stranger comes through your front door unannounced and saunters through your home." It's a shock to the system and has ramifications—anger, resentment, frustration, and a feeling of losing control. Gorillas are the same. We can start a fight by startling them, and they could unthinkingly redirect their knee-jerk reaction to a nearby group member. I have seen this happen on several occasions. And when a gorilla focuses too much of her attention on a keeper rather than on where her fellow troop members are, that can also cause serious problems. I've seen a gorilla run up and wallop another gorilla who is interacting with a keeper, which in turn sets them off and can start a skirmish. Gorilla keeping taught me to always be aware of my surroundings and how my behavior and interactions might adversely affect others.

I think about this as I take a break from the daily routine, this self-awareness we must always embrace, this respect we have for them. Truth be told, it's vital to recognize and remind ourselves that we, the keepers, are nothing more than interlopers, and in other circumstances the gorillas would be perfectly fine and self-sufficient without us. And we are and should be humbled by that. They don't need us. They simply need us to provide what they require in order to get on with their daily lives. It's not that we don't interact with the gorillas—we do—but we don't encourage it to the point of distraction for them. We want them to be focused and involved in their troop, not obsessing about us. Rob Sutherland, former head keeper of Calgary Zoo's gorilla section, describes this situation perfectly, "The continual imposition of the human personality on the gorillas will also be a very negative factor over time and one should limit this, however well-intentioned."

On this cool spring morning, the doors are wide open at either end of the building, and a soft breeze is shifting its way through the length of the Ape House. Bongo is up on the upper level of his indoor area when he decides to amble down to the front where I am now sitting. He takes a seat, while issuing a friendly belch vocalization. "Naa-hummm," he says to me.

Nothing is going on. Most of the gorillas are outside as the day is beautiful. Bridgette is nowhere to be seen; she must be in the small outdoor enclosure along with Fossey. "How are you Bongo?" I ask. He does another quieter "naa-humm." Then I say, "How's your boy doing?" No response this time, but that's OK. I'm just pleased to sit in his presence. Bongo is relaxed. He doesn't do this that often anymore, bothering to come and have a sit-down. He's busy now with a "wife" and kid. I relish it, this just hanging out together for a few moments. It's a friendly howdy-do, touching base. Maybe it's somewhat of a momentary pause from being a busy father and protector. I don't care, whatever it is, it is quiet and solid and a gift.

Keeper Training Manual—compiled by Beth Armstrong, 2004

New Keeper Protocol

1. Always be aware of where your body is in relation to the enclosures; i.e. never lean up against the mesh, don't stand too close, etc.
2. Never approach another keeper who is working with a gorilla, i.e. giving a drink, feeding, training or interacting with a specific animal. By walking up behind a keeper, a gorilla, and/or keeper could be startled. That animal may take it out on the keeper or another nearby gorilla.
3. When walking down the back aisles, whistle, sing, or talk to yourself to alert the gorillas of your presence before you come into visual contact with them.
4. If a gorilla displays, spits, or throws feces, act as if nothing occurred. Never confront or stare at a gorilla that has done the above.
5. Be very aware of where you place cleaning utensils so they are not leaning against doors or mesh or within easy reach of the gorillas.
6. Avert your eyes when talking to a gorilla. Never stare at them.
7. Never presume to touch a gorilla unless and only when so-licited or indicated by them. Even stopping to interact with (talking to) a gorilla that is friendly may upset another group member and cause a disturbance within the group.
8. Go with your gut, if something does not feel right, like you left something unfinished or you feel funny about en-tering an enclosure, most likely your instincts are correct, go back and double-check before proceeding.
9. Do not get discouraged. It simply takes time to establish a relationship with an individual gorilla. Some take longer than others.
10. Follow the lead of more experienced keepers. Do not pre-sume to have the answers.

* * *

Cora wants to play. She is an especially sweet-natured, affable gorilla and is the very same gorilla I watched years before on TV. This is the Cora that drew me to the Columbus Zoo, and as such, I owe her a lot.

Keepers and gorillas physically interact, but it is always and only at the behest of the gorilla. Some gorillas may vocalize or turn a back, looking for a tickle and the keeper returns the favor. Or some may reach out to hold a hand or tug at the keeper to come closer for a hug. Being asked to play or give or receive a hug is the ultimate compliment, but it's a dance of trust and affection, one based on mutual respect and the gorillas always call the shots.

Presuming to touch a gorilla without clear solicitation is tantamount to the most horrendous social *faux pas*—and is dangerous. It's akin to approaching a perfect stranger on the street and presuming to touch him. Invading someone's personal space is completely unacceptable and is one of the first things we all learn as a social being, be it human or gorilla.

While writing this chapter, I again found myself thinking about who we are as humans, how we have evolved, and how we navigate through our respective lives. In recent years, our apparent addiction to social media—an oxymoron if ever there was one—continues to deeply disturb me. Humans are primates, and we learn through direct social contact with others. From a young age, we take our cues from watching how others relate to us. Do they smile or scowl? Are they calm, angry, sad, gruff, or kind? We intrinsically learn to read body language. And although we have an actual spoken language with very specific words to help us understand what is happening around us, we sometimes forget that it is the visual cues that really inform us—to danger, to love, to someone initiating play.

As young primates, we not only learn the parameters of acceptable behavior from the approval and disapproval of adults, but we also learn through play with other children, without the presence of adults. I was fortunate to have grown up in a time when we spent our time talking with our friends on daily walks home from school or heading to the pool. Or just as importantly when alone, allowing for quiet reflection. When I walked home by myself, I thought about the changing of the seasons, the sights and sounds of our neighborhood, about what may have happened that day. And I worked things out in my head.

My brothers, my friends, and I were expected to entertain ourselves. Hanging around the house was not an option—unless we were reading, which was a perfectly acceptable endeavor in our home. Our parents didn't need to actively boot us outside, although they would if they saw us lounging indoors, because we relished our unsupervised freedom. We were left to our own devices. Give us a heavy cardboard piano shipping box (there was a piano store up the street from our first home) or pieces of discarded wood with a hammer and nails taken from our garage and we were off.

Interacting with the other children in the neighborhood taught us our place in the hierarchy. Who were the kids we liked, who was funny, who was zapping too much of our energy with their high drama, who was laid-back and easy to hang with? Who was aggressive, making us uncomfortable and therefore we made sure to avoid them? Who hurt our feelings and how did one cope with that? We learned to discern, to navigate our way through social situations and then to make decisions based on those experiences.

Nowadays at coffee shops, airports, restaurants, I am dismayed as I see people passively and actively avoiding one another because they are on their phones. And, yes, I, too, am guilty of overusing my phone. It is a convenient and handy device for looking up intriguing words, new books, listening to music, and seeing photos of nieces and nephews. As a conservationist, I share a wide array of information pertaining to wildlife issues on any number of websites and Facebook pages I manage.

But what frightens me is that the most basic of primate behaviors may be subverted by our seeming unwillingness and inability to sit across the table and interact face-to-face, eye-to-eye, body language-to-body language with one another. As a child, no matter what may have happened at school that particular day, once we got home we changed into our play clothes and ran out the front screen door to join friends—there was a respite from any negative experience. Now kids can never get really get away from cruelty or unkindness. This inability to put our phones down—whether looking for approval in the number of "likes," thumbs-ups, or happy faces, or cringing at posted derogatory remarks—is surely debilitating to our inner souls,

FIGURE 29.1. Beth playing with Fossey

our psychological well-being. In a gorilla's world, they may have an altercation one moment with another troop member but once the dust-up is over, it's done and everyone moves on. It happened and may subtly affect behaviors, but it is not all consuming.

And I ask this simple question, "Why do young children or even teens need a phone?" What is heartening is that finally people are starting to see the damage that is being done, that studies and reports are beginning to show that this generation of children is nervous all the time, engulfed by anxiety. They are on edge and unsettled. Endless outside stimulus combined with overscheduled kids who have little or no time to play outdoors or daydream or just mindlessly putter about is a recipe for social disaster.

As I write this, another school shooting has occurred. I have thought for years that there are many reasons why this is happening. Social media has allowed anonymity to become pervasive. It allows humans to say cruel and mean-spirited things to each other without direct social sanctions from one another and without face-to-face ramifications. Social media also allows people to obsess over unhealthy interests, such as glorifying guns or violence, or to listen exclusively to those who share their own views. When a person hasn't learned how to deftly deal with life's social difficulties, it allows for something we seem to see quite frequently nowadays, an "I am the victim" mentality. If you are the wronged party (the victim), then you can't be held accountable for cruel and unkind antisocial behaviors that you perpetrate—or so says the aggrieved person. That is when we see the threads of society, the very fabric that holds us together as a primate community, begin to unravel. Possibly it will unravel completely in the future; I hope not. But basic social norms must be learned and adhered to and we are not doing a particularly good job of it at the moment.

I was struck by something I read several years ago. Buddhist monk, photographer, and author Matthicu Ricard wrote a book called *Altruism: The Power of Compassion to Change Yourself and the World.* Ricard cited a study from a decade ago that said over 50 percent of adults under the age of twenty-five wanted to be famous—they didn't care for what—they just wanted to be famous. There is something off-kilter with that, something so completely self-absorbed, so unsocial, and so very un-primate like.

30

BRIDGETTE, BONGO, AND FOSSEY

They continue to enchant us, this threesome. As I enter from the outdoor area, I hear deep continuous laughter. Rooms 3, 4, and 5 are empty as Mumbah's troop is outside, so the sounds must be from Bongo and Bridgette. Walking up to the front of their enclosure, Bridgette is sitting snug in one of the large black rubber food tubs. We don't use them for food but rather as an enrichment item. Sometimes youngsters play in them or just drag or throw them around.

Bongo's rump is in the air, and Bridgette is batting his hands away as he leans over in order to tickle her. She is laughing so hard she can't catch her breath, but Bongo is relentless and Bridgette, with her sizable girth, is stuck. She's flailing about so intensely that the tub falls over but as her bottom is firmly wedged, she still can't extricate herself. She looks like a Weeble doll rolling around on the floor. She's getting desperate but finally is able to free herself before quickly waddling away from Bongo, and he lets roll with a contented and self-congratulatory vocalization.

Several months before, Nancy was taking photos in the morning. Fossey wandered over to his dad as Nancy held her breath. At this young age, Bridgette routinely did not allow Fossey independent access to his father. But that day, both Bongo and Bridgette were hunkered down in deep beds of hay quietly and determinedly looking for scattered food when Fossey approached his father. With Bridgette's

FIGURE 30.1. Bongo stroking Fossey's nose, Bridgette in background

body blurred in the background, Bongo reached out his enormous finger to gently touch his son's nose while Fossey looked up at him.

Later when Nancy and I had a discussion about that particular photo, she said that she raced out of the building to go get the film developed immediately—and only when she finally was able to see the photograph was she able to breathe again. She had captured something iconic. This image would appear on the cover of the national zoo association's magazine later that year.

Fos is a happy kid, confident in his place in the world, not arrogant, just assured because of the nurturing love of his parents. Currently, Fossey is scampering up and down the stoop, using his mother's body as a ladder and her extended arm that is conveniently resting on the stoop as a bridge. The stoop is the big step-up to the overhead transfer chute, but it also has a fairly large square shelf space to play on.

His dad, Bongo, sits on the other side of the stoop facing Bridgette. Fossey grabs onto his mom's arm, swings a bit before scrambling up, using it to cross over to the stoop where he then quickly jumps off, does a roll in the hay, and starts up again—up, jump, roll, up jump, roll. His dad vocalizes to him every once in a while, a bit of encouragement or approval or perhaps both. Sometimes his mother grabs Fos as he whizzes by for a quick tickle—a mere momentary delay to his routine.

Several of us are just hanging out watching the gorillas. Charlene and Nancy are both sitting on the front bench. I am on the cement floor, my feet in the low gutter trough in front of Bongo and Bridgette's enclosure. We talk among ourselves, laughing occasionally at Fossey's antics. A variety of enrichment items are scattered about the enclosure. Bongo has spied a small yellow plastic boomer ball. It's a perfect fit for his hand. He knocks it against the metal stoop, making a dull clacking sound, and then he glances over at us and tosses the ball in the air, catching it nonchalantly before turning back to us to gauge our reaction. We crack up laughing, saying, "Nice one, Bongo!" He rumbles at us and then amps it up, tossing the ball again but this time expertly catching it while looking directly at us—not the ball—making us whoop even more. Bongo has this particular look he gets when he's in a goofy mood. His eyes have a hint of mischievousness; he vocalizes with a playful "ha" and does these incredibly long rumbling vocalizations when he succeeds in getting us all to laugh. He's being funny—and he knows it. We are his rapt audience.

He continues with the ball toss. Bridgette, sitting a couple feet away, looks on in a disinterested, "I'm not impressed" sort of way. We watch, giggling when he does a particularly excellent toss, but then he stops, looks intently across at Bridgette, looks at the ball, and then seems to consider. Growing silent, we watch as his thoughts play across his face—we see it coming before Bridge does. He slyly tosses the ball once more for effect, catches it effortlessly, and in one fluid movement leans slowly forward and bonks Bridgette roundly on her head with it. She is so startled she loses her balance momentarily. Bongo, on the other hand, thinks it's hysterical, placing his hand in his mouth looking like a guilty kid caught in the act. He glances back at us. We are hooting at this point, and Bongo is rumbling and rumbling, seemingly deeply satisfied with himself.

31

WHAT'S IN A NAME?

She is the twelfth infant born since I have been in the department, and she is named Jumoke, Swahili for "everyone loves the child," "beloved child," or in the Yoruba language, "spoilt child." The latter definition will prove prescient.

I believe in the power of names. I am named for my great-grandfather, Royden Lynn Shuler, beloved father of the eight Shuler girls of Barrington, New Jersey, who passed away one week before my birth. My older brother is named for our maternal grandfather, another brother for a German immigrant great-grandfather, and my youngest brother is named for the very same veterinarian, Dr. James Hugenberg, who took me on animal rounds when I was a small child. What's in a name? They are stories, ancestors, circumstances, a reference point, a reminder, a past love, a proud connection, a web that binds us.

We have used a plethora of names for our baby gorillas. Some are named for their ancestors. Mac, one of the twins, is named for Baron Macombo (his paternal great-grandfather); his twin brother, Mosuba, is named for our nursery volunteers, Molly, Su, and Barb. Nkosi ("leader" in Swahili) was named for our ob-gyn who consulted with us during births. Binti Jua, which means "daughter of Sunshine" in Swahili, was named for her father, Sunshine. Some babies were named for people who were dedicated to saving wild gorillas, as is the case of Fossey. They all have their own story mingled and tangled up with those they are named for.

I walk around the outside perimeter of Jumoke's enclosure where she and her exhausted mother, Toni, are resting, a clipboard pulled tight against my chest with one hand, the observation notes firmly attached—notes that will guide us as to how we will proceed—and a flashlight in my other hand. Only a few minutes after her birth and I am already marveling at what a beautiful baby she is.

All newborns have that slightly bewildered look, squinting at the harsh world they now find themselves in. They are seemingly trying to get a handle on their new and mystifying environs, their heads slightly herky-jerky and wobbly on their scrawny necks. They are old-mannish looking with wrinkles around their toothless puckered mouths, wrinkles under their eyes and along their necks. Their abdomens are flat now but once they have nursed we will look for telltale signs of a taut, rounded belly. A baby that is thriving has a bit of a potbelly. Jumoke has the usual white spot on her bum that all infants have but will have faded several years from now. This spot is purported to be a visual aid or cue for the mother and other troop members to keep an eye on the infant, but truthfully, I'm not sure anyone knows what the real purpose is, as mothers are pretty adept at keeping tabs on their kids. Others have speculated that it acts as a "get out of jail" pass for the infants—upping the already high tolerance level from lenient adults. Jumoke also has the telltale swirly pink marbling on her hands and feet, with some fingers and toes completely pink. We know this unique pigmentation is a manifestation of her father Sunshine's genes. Every single one of his kids has these markings in some form or another.

While I was writing this chapter, a photo was posted on Facebook of a close-up of a sleeping newborn gorilla on its mother's chest. All you can see is the infant's side-turned face completely relaxed in deep sleep, every wrinkle on its neck visible, the slightly upturned mouth looking like a subtle smile. Most expressive are the mother's huge fingers and thumb hovering just above her baby's head. She has either just stroked the baby's face or is getting ready to. The image is evocative of pure gentleness. In those lingering fingertips, one can imagine the love and surely delight that this mother feels for her newly born infant.

Gestation for a gorilla is a little over eight months, somewhere in the range of 257 days—just shy of the 280 days for humans. We are so similar to them that it boggles the mind sometimes. Some studies have shown that we humans may share as much as 98 percent of our genes with gorillas. Gorillas mirror us in so many ways, so is it any wonder that they fascinate us?

The keepers collect urine frequently to test for pregnancy using standard human pregnancy tests. There's that indisputable similarity again. We carefully keep notations on our daily records of any and all breedings observed in the past weeks or months. When we get a positive pregnancy test, we backtrack to when we think she may have conceived, then plan the birth watch accordingly. As we get closer to the projected birth date, remote cameras are installed in the Ape House, and a room is set up offsite for the volunteers, who will begin monitoring the gorillas during the night. Docents are trained in what to look for; subtle signs of an impending birth are written down and posted: restlessness, antsy behavior, building an elaborate nest. A contact list is placed near the phone: head keeper, curator, veterinarian, etc.

Nkosi, or Nik as we frequently call him, grandson to Colo and Bongo, son to Toni and Sunshine, was adopted by our surrogate mother Bathsheba and raised in Mumbah's surrogate troop. Eventually he was moved to the North Carolina Zoo. Nik's keeper, Aaron Jesue at North Carolina, simply adored him. That affection was evident in Aaron's voice when I spoke to him recently. Gorillas do that to a person. Through their personality, they create some deep and abiding connection with us—a bond once made, never really broken. Aaron wrote several articles for *Gorilla Gazette* in 2011 and 2013 about Nik, describing his first impressions of him. "Seeing Nkosi in quarantine, we could see he was much more mild mannered and much smaller than the other boys (adult males) who continuously kept us on our feet. Little did we know, Nkosi's mild-mannered behavior would make him an instant star and later one of the most beloved animals at our zoo." Aaron went on to describe Nik's manner when introducing him to their females. "The introduction couldn't have gone smoother or quicker! We were able to proceed with ease and speed due to the accommodating nature of Nik towards his new females, while also being dominant without the use of excessive force. He really seemed ready for the responsibilities of leading a group. He respected his females, and his presence was enough to deter any issues from arising." Nik sired two infants at North Carolina, Bomassa and Apollo. There is a remarkable photo that accompanies the article with Nik's infant son Apollo riding high on his back, arms clinging to his father's thick neck—something rarely, if ever, seen with a silverback and infant. Nkosi lived up to his name—leader.

But Nkosi passed away suddenly at the age of twenty-one—a devastating blow for both their gorilla troop of three females and two infants but for their staff as well. Aaron wrote, "To replace a soul like Nkosi will be impossible. We also shouldn't ever try. The memory of his life will be carried on through Bomassa and Apollo for as long as they live and at whatever institution they will call home. The impact he has made on the other gorillas, the staff that worked with him every day, and the visitors that may have only seen him briefly on exhibit, is immeasurable. Our stories about Nik will carry on the legacy of the gentle giant that was the true leader of the North Carolina Zoo."

Interestingly enough the silverback recommended to replace Nkosi was none other than Mosuba, Mac's twin brother who had been moved from Omaha's Henry Doorly Zoo to North Carolina. Nkosi and Mosuba were biological cousins and both shared Mumbah as their adoptive father. They were raised in the surrogate group and were used to many new members being added and born into it. They were flexible, laid-back, and adept at change, thanks to their mutual upbringing. I was beyond happy to see Mosuba with a troop of his own and equally happy that Nkosi's two sons would benefit from being raised by him.

The scattering of these babies throughout the zoo community, who are now adult mothers and fathers whether biological or adoptive, began at Columbus on a night not so dissimilar to this night of Jumoke's birth.

32

EVERYONE HAS A STORY

Sylvia is a large female with a palsy-like shake of her right hand, accompanied by a facial tremor when especially nervous. But today, surrounded by a large pile of bright yellow forsythia branches, she is relaxed and happily munching away while her curious troop members look on.

Sylvia has come a long way since arriving at Columbus in late 1986. She was a part of a wave of females that were sent to Columbus in the mid-1980s in the hopes that they could be socialized. Although owned by the Baltimore Zoo, she had been housed at the National Zoo in DC for the previous ten years. Sylvia arrived in Columbus, accompanied by her keepers from both zoos and Baltimore Zoo's veterinarian, Dr. Mike Cranfield. Mike shared with me his thoughts on Syl. He said they didn't care if Sly ever bred; they just wanted her in a place where she had an opportunity to become a vital member of a troop. I was struck by the pure unselfishness of it all.

Eventually, Sylvia did breed, although she never became pregnant. But she would play an integral role in the development of our Surrogate Gorilla Program, becoming an adoptive mother and raising infants, Jumoke, Nia, and Little Joe as if her own. As with all gorillas, Syl had her own idiosyncrasies, her unique quirks. She had an unorthodox way of carrying her adopted infants. Most gorilla mothers carry their infants on their backs as they get older, but Sylvia simply wouldn't allow it. Her adopted infants learned to cling vertically

FIGURE 32.1. Sylvia and Jumoke

to her arm with their bum safely cupped in Sylvia's hand while she walked, resulting in a ball and chain sort of gait—or the youngsters learned to walk in tandem under Sylvia's belly, step for step with their adoptive mom.

And Sylvia developed a different mode of transport for the infants when transferring via the hard metal stairs into the overhead chute system. She tightly grasped the infant between their elbow and shoulder and lifted them up or down each step. Either Sylvia was oblivious to her surroundings or had a poor sense of depth perception, as it was not unusual to hear a resounding "thump" followed by a distressed yelp during transport. The infant soon learned to tighten up their body into a ball so no loose ends (arms, legs, hands, or feet) could get accidently bumped while in transit. The keepers winced more than a few times when an infant's head was knocked against unforgiving metal as a slightly perplexed Sylvia looked down at the protesting infant.

In 1987, at the age of twenty-four, Sylvia was first exposed to an infant when her troop mate, Pongi, gave birth to a son, Colbi. Both females were housed with Oscar. Pongi and Oscar were the second gorilla family to raise their own offspring at Columbus Zoo, just a little over a year after Fossey's birth. It was wonderful to see them interacting and caring for their son, but Sylvia was also becoming an essential part of Colbi's daily life as an auntie figure and constant playmate to him. If mom or dad didn't want to play, Aunt Syl could be relied on for a go.

Eventually Syl was moved to Bongo's group, and then to Mumbah's surrogate troop. In 1991 she became an adoptive mother to our nursery-reared infant, Jumoke. Due to several factors in the department at that time—a gorilla death, an introduction, as well as another birth—we had to delay Jumoke and Sylvia's introduction. This, unfortunately, allowed Jumoke to become even more attached to her human caregivers.

The adult's reaction is our overriding concern with any introduction involving a large adult gorilla to a much smaller infant. Injury to the infant is always a very real possibility, whether intentional or, more likely, accidental. We are always apprehensive on the day of opening the door between an infant and an adult.

Our behavioral observations prior to the introduction indicated that Sylvia was very interested in Jumoke. She and Jumoke sat across from one another consistently at the mesh between their enclosures, interacting frequently, doing the usual: sharing food, toys, and blankets. These behaviors suggested that Syl was inclined to be gentle with Jumoke. And a key observation: Sylvia was reluctant to join her group outside if Jumoke was spending the day inside.

Gorilla life revolves around very basic social norms, as do our human lives. Adult gorillas are incredibly tolerant and will put up with a lot, but there are certain things you just don't do as an infant or juvenile. Overtly aggressive behavior directed at an adult from a youngster will be countenanced for about a nanosecond, just long enough for the adult to react with a "you did not?" expression on their face. A youngster could be in a world of hurt if they persist with such unsuitable behavior.

The keeper staff set up the building for the introduction to begin, cleaning early and adding extra hay and distributing a variety of treats. With one last affirmative collective nod of our heads, we opened the front and back doors between their adjacent rooms. Jumoke immediately hurled herself at Sylvia—think atomic rocket—chasing Sylvia from one end of the enclosure to the next. This eighteen-month-old infant, weighing in at well less than a third of Sylvia's 190 pounds, was in an absolute relentless frenzy, cough-grunting, lunging at, and trying to bite Syl.

According to gorilla social rules, Syl had every right to nail Jumoke, to bite her, cuff her, or push her away, but instead Sylvia just kept backing up and when necessary, peeled Jumoke's irate body off her own arm, leg, or whatever appendage Jumoke happened to have attached herself to. This went on for some time until Jumoke finally settled down. But not before Jumoke aptly earned the nickname "Velcro."

We were stunned, mouths open. But once the initial shock wore off, we began praising Syl for her patient forbearance. Sylvia continued acting in a nonchalant manner, never approaching Jumoke, instead allowing Jumoke the time and space to figure things out. If and when positive contact was made, Jumoke would have to make the first move.

This allowing of space, with a calculated disinterest, is one of the smartest tricks in the primate social handbook; I have seen a male gorilla employ this tactic with skittish females they wanted to breed with. I also saw this behavior when a male woolly monkey engaged in a deliberate nonchalance when a very young and very frightened female was introduced to him. He went so far as to lie on his back, his arms and legs fully extended. He made himself completely vulnerable and sure enough, the female woolly felt safe enough to approach. And then, he gently touched her with his outstretched hand. This tactic is very similar to when you turn your back on a horse, not forcing an interaction. In most cases, their curiosity overcomes their fear and they approach. Sylvia and Jumoke worked out their differences, and Sylvia became one of our premier surrogate mothers.

I paid a visit to the National Zoo five months after Sylvia was transferred to us. While being given a tour of their gorilla building, I noticed large water-filled plastic barrels holding an abundant amount of blooming forsythia branches. Many North American zoos were just beginning to understand the importance of giving trees branches, such as willow or Bradford pear, to gorillas for an activity as well as for digestion, but forsythia was a new one on me. I asked their head keeper, Melanie Bond, about it, and she told me their gorillas absolutely loved forsythia in the early spring when it was in full bloom. Then Melanie casually mentioned that Sylvia in particular went nuts over it.

Upon my return to Columbus, I talked to the other keepers, and we agreed to give forsythia a try. We scouted the zoo grounds and cut a large amount of forsythia, placing it in the outdoor gorilla yard, and then let Sylvia's troop out. Sure enough, as soon as Syl saw the yellow blooms, she started vocalizing loudly and plopped herself right down amongst them. She picked up a branch, placed the bottom end between the well of her thumb and index finger, and moved her hand up the branch, expertly and efficiently stripping it of all flowers in one smooth move, stuffing all the blooms in her mouth at once. She was a forsythia-eating machine. Her curious troop members gathered around watching her. Within several days, all the other troop members were eating forsythia with the same gusto.

33

LULU—A GAME CHANGER

Lulu came to us in 1984. She had been wild-caught in what was then known as Equatorial Africa, sometime around 1964, and taken to the Central Park Zoo (one of several zoos that made up the consortium of New York City zoological institutions known as the Wildlife Conservation Society, which included the Bronx Zoo where WCS was based, Prospect Park Zoo, and Queens Zoo) where she resided for twenty years until her arrival in Columbus. While at Central Park, she gave birth to two infants, one of whom had died and the second, named Patty Cake, she had raised. Lulu had a petite figure, a sweet face, and a habit of sticking the tip of her tongue out. Some of us called her Boo, which she always answered back to. What I really admired about Lu was she was not needy of our attention and affections but was kind enough to let us into her world. Over the years, two personality traits became apparent to the staff: (1) under no circumstances would she exhibit submissiveness to a male even though her life might have been easier if she had, and (2) she had a nurturing maternal quality that was an innate part of her. She was at heart a natural mom.

In February 1987, after a long labor, Lu gave birth to a female infant. Unfortunately, Columbus Zoo had agreed to a breeding loan contract that was the norm at the time with the Wildlife Conservation Society. The first infant was owned by the dam's (mother) zoo, and they wanted the baby pulled. Because this was a three-way breeding loan, the second birth would go to the sire's (Sunshine) zoo, and the third to Columbus, as we were only the host zoo, the matchmaker zoo. Indicative of the times, the Bronx Zoo (where their gorilla nursery was based), like so many other zoos, had a long history of pulling infants after birth to hand rear them. Having watched Lu go through a difficult labor, it was heart wrenching to then see her anesthetized and have the infant taken from her. The Ape House staff was vehement in its concern that this situation could never happen again, and that in the future, all subsequent agreements with other zoos would clearly state that regardless of "ownership" of the infant, the mother would be given every opportunity to raise her own infant.

Lulu was housed with the young male Sunshine and female Toni. Sunny was still very immature. As happens with blackbacks (who are in essence still juveniles—older juveniles for sure but juveniles nonetheless), he quite possibly felt the need to exert his authority over his females, something that never sat well with Lu. And as he was quite fond of Toni, she got a pass. Lulu did not.

The following year on Saint Patrick's Day, Lulu gave birth to another female infant we named Binti Jua. Right from the start, Lu appeared ill at ease, frequently pacing or placing the infant under her bum and genitals and then bouncing on it. Her anxiety when around Sunshine only exasperated the situation. She never seemed able to settle down.

Binti was pulled several weeks after her birth due to what appeared to be an inflamed eye. She was weighed, examined, and the eye treated—a piece of hay had wedged itself under her eyelid. She was given back to Lulu, and Lu continued to care for her, but still exhibited jumpy behaviors, never seeming to truly relax in her home troop.

In the next few weeks, the infant did not appear to be gaining weight. I didn't feel that we had a handle on what was going on with Binti or Lulu or the troop for that matter. My concerns were noted sporadically on the daily Ape House records, while other keeper's notations indicated that all looked good, but I couldn't shake the feeling that something was not quite right. I became quite vocal and pushed for the baby to be pulled, which was completely at odds with what I believed and had fought against for years. At two months of age, Binti was pulled for another physical exam, and we discovered she had only gained one ounce since her eye examination five weeks before.

As Binti belonged to the San Francisco Zoo, and they had been fine with us leaving her in with Lu from the beginning, I'm not sure why we didn't try and integrate Binti back in with one of our other groups when she was a bit older and her weight had increased. In the end, she was shipped back to San Francisco and ultimately ended up at the Brookfield Zoo in Chicago where she eventually gave birth to her own infants and raised them.

Most people will remember Binti Jua as the gorilla who in 1996 gently cradled the three-year-old boy who had fallen into the Brookfield Zoo's moated exhibit and was knocked unconscious. Binti went over and carried the boy to the closed transfer door that led to her back holding area. She sat with the boy tenderly draped over her lap, patting him in what I can only interpret as comfort.

The media storm that ensued was typical, reporters sensationalizing the gentleness of this female gorilla as if it were the greatest surprise on earth. My experience with the media concerning gorillas is that they have consistently missed the mark, describing normal gorilla behavior as if it is unusual or out of the ordinary. Newspaper accounts attributed Binti's tenderness to having been raised in a nursery setting, insinuating that her maternal behaviors could have come only from the influence of humans, as if only humans are capable of kindness and empathy. That kind of perspective is aggravating to say the least. We would do well as a species to take in what gorillas have to teach us in terms of their parental behaviors.

Clearly Lulu had a record of allowing Sunshine to breed with her, but she was clearly ill at ease in his immediate company for any prolonged period of time. Her unwillingness to submit to him infuriated Sunshine and resulted in his dogged pursuit of her. It was not unusual to see Lulu high up on the habitat's catwalk with Sunshine in relentless pursuit.

Lulu again became pregnant, and in late 1990, Adele came up with an absolutely brilliant idea: that Lu might do better in a troop where the male was not interested or particularly invested in establishing his dominance. And who had a more easygoing countenance than Mumbah? Adele thought that Lu stood a greater chance of raising her infant in Mumbah's troop, in a more relaxed setting.

Although, Mumbah had accepted unrelated infant JJ and the juvenile twins, it was an open question whether or not he would accept an already pregnant female within his troop and, after the birth, whether he would accept that infant as his own. Because infanticide (adult males killing infants that are not their own) does occur in the wild, we were cautious.

We came up with a plan based on the pros and cons and, most importantly, risks. We also analyzed what we had done wrong with Lulu and Binti. In retrospect, we recognized that we should have given her an option out of Sunshine's group when she wanted so she could relax, which would have allowed her to nurse more often and interact with her infant in a stress-free environment. But we also recognized that Sunshine still might have become overly aggressive when she was given access back into his troop, perhaps venting his frustrations on her because he was not in control. So the best solution seemed to be as Adele suggested: move Lulu to a new setting, a new group. Adele's idea set the program on a whole new groundbreaking course.

On the last day of January 1991, while in Mumbah's troop, Lulu gave birth to a daughter we named Kebi Moyo. All went well, and Mumbah gave no indication that he would be aggressive toward the infant. We keepers used to joke that poor Mumbah probably woke up on some days only to think, "WTF, where did this kid come from?" then count off troop members on his fingers, and shake his head in puzzlement at the latest addition.

Lulu was given access out of the troop if she chose, especially at night. She loved the overhead chute, which she packed nightly with hay, making a comfy sleeping nest. As a precaution and because of the lack of weight gain of Lulu's last infant, Binti, Charlene and the other keepers worked with Lulu prior to the birth, getting her familiar with a baby bottle. This would allow us to give supplemental feedings to the infant if need be. While the infant was given her bottle, Lu received one as well.

After every birth, we set up a twenty-four-hour watch lasting several days to weeks, depending on the particular female. I was taking night classes at Ohio State, so I usually was assigned the late shift, driving directly to the zoo after my class, arriving around 10 p.m.

FIGURE 33.1. Charlene giving a bottle to Lulu and Kebi

FIGURE 33.2. Lulu with her daughter Kebi, who has a play-face

I was always filled with anticipation as I drove across the hushed and pitch-black zoo grounds, parked the car, and then quietly entered the Ape House. Even after all the years I had been working with gorillas, I still had that sense of privilege, that I had somehow won a prize. I was being given the unique opportunity to observe them in such peaceful moments, especially a mom and her newborn baby gently interacting.

In 1999 Kebi gave birth to her own daughter, Kambera Dupe, via cesarean section. Although Kebi was given the opportunity to rear Kambera, she was unable to do so. Lulu stepped in and adopted her granddaughter and so became the fourth surrogate mother at Columbus. In 2006 Lulu adopted another infant, a male from Cheyenne Mountain Zoo, once again becoming a surrogate mother. All the qualities we saw in Lu when she first came to us in the 1980s were reinforced over and over again throughout her lifetime. She was and had always been a natural mother.

34

A NEW LIFE FOR BATHSHEBA

In March 1988 I flew to Colorado to pick up a thirty-one-year-old female gorilla named Bathsheba. Born in 1957, she was one of many from her generation of gorillas who had been captured from the wild in the 1950s and 1960s. Apparently, she did not fit in with the other gorillas at her zoo—a frequent refrain we heard in connection with many gorillas.

When I first laid eyes on Bathsheba, she was in an enclosure that I can only describe as a cave, a man-made cave with three solid walls with a barred front. She was living in a basement-like setting with little or no light, and her exhibit was located below the other gorilla enclosure. It was so dark I had a difficult time seeing her. This is not an indictment of that particular zoo; at the time we all were way behind the curve in providing proper housing for gorillas. Zoos were hovering on the cusp of seismic changes that were long overdue, but a recognition that things had to change was beginning to percolate.

While I was there, I could not shake a feeling that I was being managed, that the keeper staff hovering in the background were hesitant for some reason. I didn't think it could get any stranger, but while I was being given a tour of their nursery where they were raising an infant orangutan, their curator left the room. He came back with the perfectly preserved frozen body of another infant orangutan that had passed away some time before. I was appalled and was left completely speechless. To their credit, the keepers too looked shocked, a better word might be mortified.

Eventually I was left alone with the keeper staff, and they relaxed. But it was clear that there were more details that were not part of the official record. A keeper approached me, "I need to tell you something, Bathsheba gave birth years ago and then killed the infant. I thought you should know." I had not read any of this on Bathsheba's records but after having seen where she was housed, I thought, "Hell, who could blame her?"

When I got back to the hotel that evening, I called our curator, Don Winstel, and told him what had happened that day. I described the cave-like structure Bathsheba was living in and the death of her infant. Even with that information, Don, true to form, never balked at bringing Bathsheba back to Columbus. We knew we could provide her with a good life. We would make sure she would be given every opportunity to blend into a gorilla group. If she bred, great, if not, that was fine as well. She would live out her life in a comfortable and socially stimulating setting. For her part, Bathsheba would prove herself to be a real asset to our surrogacy program.

I was there only a couple of days before it was time to head back to Columbus. Bathsheba was lightly sedated, placed in a wooden travel crate, and loaded into a zoo vehicle for the drive to the airport two hours away. Bath and I had a few hours before leaving so we waited it out in the cargo holding area at the airport, where the airport staff was curious but incredibly respectful, keeping their distance. Gorillas have that effect on people. They make people more circumspect in their movements when around them. Bath was a sweetheart, vocalizing softly to me when I offered her food, no displays, no cough-grunting, and no drubbing of the crate.

The fantastic thing about this particular airline was that I had easy access to the cargo area of the plane so I could check on her frequently during the flight to give her more food and water and make sure the temperature was warm enough. There was a small area right behind the cockpit door that had five passenger seats, behind that was the door to the area where Bathsheba was. The flight crew proved to be kind throughout the entire flight, asking frequently if Bath was comfortable, was she warm enough? Before becoming airborne, they asked if I wanted to sit up front with them during our takeoff. My response, "You bet." So there I was sitting behind the pilot and copilot, across from the navigator, looking out at the inky black night, the engines revving as we gained speed. The blue runway lights blurred on either side and we lifted off toward home as I softly hummed Joni Mitchell's "This Flight Tonight."

In late 2016 Dr. Richard Wrangham, primatologist and professor of biological anthropology at Harvard University, was in town. I go back a ways with Richard, having met him at a chimpanzee conference in 1991, thanks to our mutual friend Ann Pierce. That chance meeting resulted in the Columbus Zoo supporting Richard's Kibale Chimpanzee Project in Uganda beginning in early 1992. That project gave the Columbus Zoo an opportunity to build a conservation partnership model that was used in formulating future partnerships with other in situ projects. We allocated funds for their research at Kibale, created their logo, developed educational materials and conservation posters, made T-shirts using the same artwork for their staff, and provided rain gear and backpacks for the Ugandan field staff. We sent our veterinarian to Uganda and brought the Ugandan vet to the United States. We also provided university fees for a Ugandan master's student. All of this stemmed from Richard's and my chance meeting at a primate conference.

Here's what I know about Richard. He is a first-rate academic, highly respected, and world-renowned. He was one of the first group of students at Jane Goodall's Gombe site in Tanzania in the early 1970s. And he loves to share stories, delighting in the telling. Just as importantly, he loves listening to the experiences of others.

I joined Richard and our fellow primatologists at a local bar for drinks during his visit to Columbus. Richard began by telling me of an eight-year-old male chimpanzee named Kakama (a story I later remembered I had read in his book, *Demonic Males*) who treated a piece of wood as if it were an infant, carrying it, appearing to play with it, making and sharing a nest with it, even retrieving it when it fell from high up in a tree. Kakama was doing all the things an older sibling might do with a younger brother or sister. At that time of Richard's observation, Kakama's mother was pregnant again, and, in essence, this young male was practicing maternal-like/sibling play behaviors. The storytelling torch was then passed to me, and I shared a story of a thirty-two-year-old female gorilla that found dead birds in her outdoor enclosure on several occasions during one particular summer and proceeded to display behaviors quite similar to the chimp Kakama.

Remember that wonderful scene in the Harry Potter movie when Mr. Weasley finds out that the Weasley boys have taken the flying blue car out for an illegal drive to rescue Harry from the Dursley house? And upon finding out about it, Mr. Weasley, who is an avid collector of Muggle (non-magic humans) gadgets and is fascinated by all things Muggle, leaned in and asked the boys in an enthusiastic and genuinely interested voice, "How was it?" just before being roundly scolded by his disapproving wife. That's what sharing stories with Richard is like, it's a complete all-in, all ears, listening intently to the details, delighting to hear of such things, whatever they may be.

Bathsheba is a relatively petite gorilla, her arthritic toes and fingers perpetually curled in. She meanders along tentatively, sometimes pulling her legs up close to her torso and swinging her body through her crutch-like extended arms. The movement is similar to a pendulum, but always in a forward motion, her body's momentum carrying her along.

She has adapted well to her life here at Columbus. Bath is an integral part of the surrogate troop led by Mumbah and is sharing her life with several other adult females and youngsters. It will be a few years before she becomes a mother herself by adopting fourteen-month-old Nkosi in 1992.

But for right now, she has found the body of a dead mourning dove in the outdoor enclosure and is doing her darndest to mother it. Bath tucks the dove in the crux of her right crotch-pocket (where her upper thigh and lower abdomen meet), pulls her legs up to secure it, and moseys through the exhibit. When she stops moving, she does a thorough exam of the dove. As a keeper, it is fascinating to watch and note this careful maternal behavior from her. It mirrors what has been noted in the wild at the death of a gorilla infant. The mother will continue to carry her deceased baby for days, cradling it in her arms, placing it awkwardly on her back for transport or next to her as she settles into a sleeping nest. We watch in fascination as Bathsheba adjusts the carcass as she climbs up and down the exhibit. It's her dove we figure, so we make no effort to retrieve it. In fact, we spend most of the day hooked, watching her.

At one point, Bath is on her back near the large plate-glass windows where the public gathers to view gorillas, the dove resting on her belly. She gently lifts it into the air, stretches its wings wide apart, its head lolls grotesquely forward, and the keepers and public share a collective "ewww," but we are all so entranced that we continue to watch. Bath stretches her legs straight up and supports the bird's body with her feet as she continues to stretch the wings. It's a version of airplane we often see mother gorillas do with their infants, much to the delight of both. Bath continues until she tires of the game, then she tucks her baby bird back into her crotch pocket and moves on with her day.

35

BEHIND THE SCENES

Muke is large for a female gorilla. She's a "big-boned" girl, as we like to say. Muke has a calm demeanor and was housed with Oscar and his female in the early 1980s. As mentioned earlier, their outdoor yard is sunken, well below the public viewing area, surrounded by an even deeper dry moat. The top of the exhibit has a cement wall about five feet high that runs the circumference, surrounded by heavy hedges and then a wooden fence to keep the public at bay. From the top ledge of the public wall down to the bottom of the dry moat, it must be a good thirty-foot drop.

So it was with some surprise that early one morning, I heard a meow emanating from the outdoor exhibit. Climbing over to the perimeter wall and pushing through the thick shrubs, I peer down and sure enough, there is gray tabby, looking somewhat lost and forlorn, huddled in a corner of the dry moat. I have just let the gorillas out, and there is scatter food all over their yard. They are all happily foraging, oblivious to their visitor, so I am reluctant to bring them back in so soon. I don't want them to become agitated and possibly aggressive because of the disruption in morning routine. Because they don't seem to notice the cat, I come up with a brilliant plan, if I do say so myself.

I head into the building, and open the solid metal door leading from the yard into the building and begin calling, "Here kitty, kitty, kitty." Unbelievably, the cat actually scampers in—along with Muke.

I am standing in the keeper aisle in front of the tall barred door looking into their indoor exhibit when Muke walks over and settles down
on the other side of the door. The keeper door has an opening that
allows us to pass through food items, branches, and blankets to the
gorilla. It's an opening big enough for my entire arm to fit through,
allowing me to grab the cat—but at the moment Muke is happily sitting by the door just having a nice visit with me. I shut the door leading to the outside yard, locking both Muke and the cat in. Eventually
I see my chance as Muke moves away from me. Calling the kitty, who
comes over meowing and stands up against the barred door, as I reach
in and swiftly pull her out of the exhibit. Muke walks back over, and
we continue with our visit before I take the cat back to the main Ape
House and lock her in Dianna's office for the day. We named her
MukeToo. After my shift, the cat goes home with me, where luckily
my roommate has a friend who just happens to be looking to adopt a
cat. So MukeToo, intrepid explorer and friend to gorillas, has a new
home and her owner has a unique story to tell.

Storytelling is a much-loved characteristic among all gorilla keepers.
Some stories stay within the boundaries of the zoo, and others get out
there in the wider world because of their significance. When we come
together after our respective days off, we inevitably swap stories. Fellow
keeper Charlene and nursery keeper Barb Jones told this one to me.

Another gorilla infant was being raised in the nursery but was being
prepped for an early introduction back into our surrogate group. Barb
brought the baby down every day to acclimate it to the sights, sounds,
and smells of adult gorillas. Because it is summer and the weather
warm, she spreads a blanket in the keeper work area between the large
outdoor habitat and the small outside enclosure attached directly to the
building. Mumbah's troop is outside in the big habitat and is the targeted group for this baby's integration. Colo is spending a good deal
of time staying close, watching the baby. Scattered on the baby blanket
are a variety of items to keep the infant sufficiently occupied, including
a plastic key ring with five large, dangly pastel-colored keys, exactly the
kind you might see in a human baby's playpen.

Colo, ever the patient observer of human foibles, sees her chance
when Barb takes the baby inside to heat up its bottle. Colo leaves

her observation deck where she has been perched watching the infant and schlepps off through the exhibit—apparently in search of something. Once she finds what she wants, she makes her way back.

She now has a long tree branch. She strips it of the smaller protruding limbs and leaves and proceeds to thread it through the mesh. She is fishing for the baby keys. Colo works methodically, stretching her arm as far as she can, stops, and seems to work out ratios and distance before she repositions herself for another go. She succeeds in threading the key ring with the branch and starts to gently reel it in, ever closer, before it drops off the branch. Shifting her position again, she finally securely snags the keys after several attempts and is able to pull them close enough for her to grab with her fingers and bring them into the exhibit.

Barb comes back out with the baby, and Colo just placidly looks at her with an innocent expression. When Barb realizes the keys are gone, she goes back inside to tell Charlene what has happened. Charlene, armed with peanuts, comes out to barter: treats for the key ring. Colo tantalizingly gives Charlene a glimpse of the key ring, which she has stealthily hidden under her foot, but it is clear peanuts are not going to cut it. So Charlene heads back inside, rummages through the fridge, and brings out fresh pineapple pieces, which Colo deems sufficient to begin the bartering process.

Charlene puts her hand out for the key ring, and Colo, after what can only be interpreted as careful consideration, breaks a key off the ring and then breaks the key in half and hands the broken half to Charlene. Colo, a savvy gorilla, knows an opportunity when she sees it so she looks at Charlene, then looks at the remaining keys and continues to barter for the best deal she can get for the merchandise in her possession. And so it goes for the next fifteen minutes, Colo breaking each key in half, handing them over bit by bit until she has ten yummy pieces of pineapple in her hands, and Charlene has ten plastic bits of keys in hers. Mission accomplished.

36

MONUMENTAL CHANGES

It is 11 p.m. when I approach the building, and I am apprehensive as to what I might find. Walking down the darkened back aisle, I pass the enclosure adjacent to Bridgette's where Bongo is housed. Glancing in, I see their fourteen-month-old son leaning against his father's massive silver-black leg, and an immediate alarm bell goes off in my head. I mutter, "Oh lord, we're in trouble."

Coming around to Bridgette's area, I see her rotund belly moving rapidly. Calling back to the volunteer who was doing an all-night birth watch on another gorilla, I say "Martha, give me fifteen seconds, I need to get Bridgette's respirations." Martha hears the concern in my voice and gives me the time I need before asking, "How's she doing?" Bridgette's respirations were high.

Approaching the front of the enclosure in order to get a better view of her overall condition, I see her standing and biting the hard mesh, as if to transfer her pain to it. She lets out a cry. Shortly after, her son, Fossey, approaches from his father's adjacent room through the baby-size door. He wants to nurse, but Bridgette pushes him away. Fossey whimpers and returns to the comfort of his father and I am left utterly and thoroughly stunned.

My mind drifts back to several months prior, when I watched Bridgette quietly interact with her son. He was in an ornery mood, full of energy, and pestering his mother while she tried to rest. They were together in a hay nest in the relatively confined space of the chute. Fossey was swinging from the ceiling mesh, using his forward momentum as a battering ram to kick his mama in the face. He was thoroughly enjoying himself. Scooping his tiny bum in her large hand, she gently pulled Fossey toward herself, holding him tightly against her chest, effectively trapping him. She began to tickle him until her laughter and his became one. My eyes had teared. How lucky I was to witness such an intimate and gentle way to discipline an infant.

I come back to the present and watch as Fossey approaches his mother once again, but she cough-grunts him away. Bongo then cough-grunts at Bridgette as if to correct her unexpected behavior, but she appears oblivious as Fossey forlornly returns to his father, this time holding his right arm across his chest, a reliable sign of insecurity and uncertainty for him. After that, Bridgette lies back down and appears to be resting. One hour later, Fossey returns to her and attempts to nurse. She pushes him away and this time something breaks in him. Fossey climbs on her back and begins screaming in frustration—his lips curled back, teeth showing—and beats her with his fists. Most alarming of all, Bridgette does nothing to dissuade him. A moment later, Fossey, softly hooting, with his arm across his chest as if in a half hug, resignedly climbs down from her back and goes back to his father.

My shock turned to alarm. I had never seen her ignore her son and have never seen Fossey pitch a fit, a behavior unusual in a gorilla. It was then that my worst fears were realized, that what we were facing was the very real possibility of Bridgette's death.

My mind turned to her mate, Bongo. Bongo, who is the most handsome of gorillas, who always has the freshest breath, a citrusy smell, Bongo who had made my life difficult when I first began working with gorillas, who had been housed on public display for decades and never able to get away from the oftentimes insensitive and jeering public. For him, a proud and regal animal, his life had been a living nightmare.

Eventually, Bongo was given this renovated home where he never had to face the public again, if he chose not to; he had more than paid his dues. In addition, he was introduced to overweight,

even-natured Bridgette. They were a good match, both in their late
twenties and both seemingly in tune to one another's moods. They
reminded me of an older couple who find each other late in life and
realize and appreciate the miracle of finally finding their respective
mate. Bridgette gave birth to Fossey when she was twenty-five and
Bongo thirty. I could not wrap my head around the idea of a future
without Bridgette's presence in both her son and Bongo's lives. Another keeper arrived and I am jarred back into the present. By now it
was 2 a.m., and after explaining my concerns, I leave to go home.

Having slept badly, I arrive back at work early the following
morning. Charlene shares her concerns with me; now Bridgette is no
longer eating or drinking. The vets come by to check on her and then
move on to their morning rounds. The other keepers leave to deal
with the daily feeding and cleaning of the remaining gorillas. Charlene and I stay to tend to Bridgette. During the night, Fossey's "baby"
door had been closed, barring access to his mother. He and Bongo sit
quietly watching her.

At 10:55 a.m. I begin having difficulty charting Bridgette's respirations. Trying to keep the panic out of my voice, I call, "Charlene,
you'd better call the vets, I can't seem to get her breathing down." As
soon as Charlene hangs up the phone, the words "She's not breathing"
come unwanted out of my mouth. I grab the walkie-talkie, asking
Doc to return to the Ape House immediately. We open the door to
go to her and begin chest compressions, as I mutter under my breath,
"Come on, Bridge, just breathe." Charlene and I both watch as the
color of her gums fades from a healthy pink to gray. When the vets
arrive in less than two minutes, they too attempt to resuscitate her.

The quiet in the Ape House is palpable; an eerie respectful silence
permeates the air. I cannot bear to look at Bongo and his son; both
are so absolutely still and unmoving. Only when we move Bridgette's
body out of the building does Bongo react. He stands on the stoop
leading up to the overhead chute, indicating for me to open the door
for his access. Then he walks all the way to the end, enabling him to
peer out the building's side door where Bridgette's body was being
carried to the waiting zoo van. As I watch, I feel as if my heart would
break for him.

Bongo and Fossey went through a harsh adjustment period for
several weeks following her death. We heard Bongo's high reverberat-

FIGURE 36.1. Fossey leaning up against his dad Bongo

FIGURE 36.2. Bongo and Fossey in the habitat

ing mourning call, a vocalization specific to gorillas, which indicates great sorrow. No human words can express grief as clearly and succinctly as that call. Fossey had to learn that he could not climb on his father's back as he had his mother's, so he quickly adapted, settling for leaning against his father's ample bulk for comfort. Shortly after her death, Bongo began making a hay nest for his young son directly beside him each night, so they could sleep together, giving Fossey the security a young gorilla thrives on.

Six months after Bridgette's death, Fossey became seriously ill, but Bongo tended him through his illness, cupping his enormous hands around Fossey, softly vocalizing to him when he was beset by cramps. It was during those few days when the very idea of Bongo possibly losing Fossey was debilitating to us in the Ape House. We could not, would not, allow ourselves to even consider it, knowing it would break him. Fossey recovered, much to our communal relief.

It is absolutely gorgeous outside. Fall has come early this year. It's been a wet one so the grass is at its greenest in the gorilla yards—similar to the deep verdant green of Ireland. Red, orange, and yellow leaves gather and flutter across zoo grounds and gather in colorful piles along the gorilla habitat's low cement wall. The air has autumn's peculiar coolness to it, and the light has a hint of gold and throws long blue-black shadows across the zoo grounds. It's one of those days that just bowls one over—the smells and the light, as if a form of magic has descended, beguiling a person with the beauty of an autumnal day. Today is dry, so the air smells distinctly of desiccated leaves and I hear their crunch as I walk from one building to the next.

My fellow keepers and I are in a state of shock, stunned by the death of Bridgette the day before, but we stick to routine, getting Mumbah's troop outside to enjoy this incongruous glorious day. We worry about Fossey and Bongo, about their loss, about their deep sadness. Bongo's mournful calls drift out to us while we shift animals outside, and it feels as if a knife is cutting me, us, as if our hearts are breaking again and again with every echoing note.

We keep a close eye on Bongo and Fossey, documenting their interactions. The keeper staff is not for one moment concerned that

FIGURE 36.3. Pongi and her son Colbi

Bongo won't take care of Fossey; we're just keeping tabs on how we can soften the blow, anything to keep them occupied. Fossey has adapted immediately to drinking his milk from a cup twice a day, while his dad receives a treat drink simultaneously. He settles next to his father, leaning into him constantly. He is comforted by his father's close proximity, and he follows him everywhere. They are inseparable, these two, one motherless and one a bereaved mate.

Life has a strange way of bringing us back, displaying timely miracles in the midst of the deepest sadness. In the north building, Oscar's mate Pongi is in labor. So we are managing two situations, consistent observations of Bongo and Fos and a team over at the north building is keeping tabs on Pongi and Oscar. We have learned from past mistakes during births. We've learned not to intervene or disrupt daily routines but, more importantly, we have learned from Fossey's birth to keep the family together. We've learned to watch closely to give the new family what they need prior to, during, and long after the birth, and all goes well. Pongi is a skilled mother to her newborn son, whom we name Colbridge, in honor of Bridgette in combination with Columbus. His nickname is Colbi. Oscar is a solicitous and gentle father, and their third group member, lovely lumbering Sylvia, is a wonderful aunt to Colbi.

Bongo and Fossey would be alone for more than a year, but eventually Bongo was given a new mate who was both tolerant and gentle with Fossey. I struggled for the longest time with my initial anger at seeing such a fine animal as Bridgette die. Regrets lingered for what I know could have been and what was taken away from Fossey and Bongo. The regret of Bridgette not being there to influence Fossey in his formative years could still at times rear its useless head. But often when I watched Fossey, when I caught him unaware, when he was totally absorbed and intent on something that intrigued him, his lower lip protruding in utter concentration so like his mom, I could clearly see his mother's face and her mannerisms in miniature. Then I knew she had left her endearing mark on this world.

37

RAINY DAYS

It is 7:30 a.m. and my dog and I are walking along the banks of the Scioto River. It is quiet, and we are alone on this early spring morning except for a funny little skunk that is trundling along the road completely oblivious to us. I am a true pluviophile, defined as "a lover of rain: someone who finds joy and peace of mind during rainy days." Everything is saturated—the soil, the road where rivulets of water course their way down to the already high river. Dampness permeates the air, clinging to everything. I hear the honking of Canada geese as they fly down the length of the river, and it is that sound, probably more than any other, that takes me back to the zoo, to the early days, to the barn, to the Children's Zoo, to walking and working along the river doing early morning rounds, hosing the polar bear exhibit, feeding cheetahs. The geese were a constant soundscape; and today their collective honks transport me back thirty-five years.

The rain has been falling for days, typical for early spring in Ohio. It is so dark and dreary as I walk across zoo grounds that it feels as if night will descend at any minute, but it is seven in the morning. The deep smell of wet soil is rising as if a vapor, and rain is plopping on the already formed puddles. My raincoat smells reminiscent of the bright yellow raincoats of grade school days, the hood an echo chamber—*dap dap dap*—as the raindrops hit my head.

Walking into the Ape House kitchen is both a relief and a welcome. Here is what I love so much about rainy days: walking into a warm, dry building, divesting myself of the layers of outdoor gear, and feeling snuggly and warm, as if coming home. The Ape House has that very feel on days such as this, and not just because of the inclement weather but because the storm will keep people at bay. Because it's raining, the gorillas will have a day of rest, free from the visiting public. Even those gorillas that have public viewing windows will not have to deal with people, as so few will venture out on a day like this. It's tranquil in the building and is one of those illusive days of grace, where everything runs smoothly. The day feels unhurried and calm as if we are in some space machine, isolated in our solitude.

After the morning routine is completed, we take a seat to do some observations. There is a playful adult male doing a twirly in the middle of his enclosure. Eyes closed tight, round and round he goes, he is four hundred pounds of sheer revolving joy. He opens his eyes only when he stops to get his bearing and find his balance. In another enclosure, a juvenile chases a youngster. They move so fast up into the interconnected chute system and back down into the adjacent connecting enclosure that it's hard to keep track of them. It started off as fun, but now there is an element of intensity. The smaller one is getting spooked so she heads straight to the nearest adult to hide behind for protection. The older juvenile stands nearby hoping the youngster makes a run for it but to no avail, so he saunters off looking for trouble elsewhere. In the chute, adult female Lulu is building a nest—again—she does love her nests. Mumbah is in a pile of wood wool resting, something he is quite proficient at. Throughout the building, we hear the murmurs of laughter from gorillas at play.

It is still raining. I can hear the steady pattering on the roof. Earlier, there was a tremendous downpour, its intensity sounding as if a train was coming through the building, but now it is just a methodical thrumming. I head to the kitchen to make a new pot of coffee. Dianna is already there making a big pot of lentil soup for the gorillas, just because. It's something different and good for them. I'm not sure they really like it, but they give it a try when served.

All of a sudden, we hear whimpering and then screaming. It is one of the twins, and he has slipped his arm through the bars of his back door, and his arm is now tightly wedged. It is a flaw in the design of the door, which allows him to easily place his arm through at the very top and slide it down, but as he does the space between the bars narrows and now he is caught just above the elbow, and he is panicking. It is similar to a Chinese finger puzzle—the more you pull away the tighter the vise grip.

This is easily fixable. We just need to push his arm up to the wider space, and he can withdraw it. But at this point he's so frightened that the other troop members have picked up on it, and they're getting nervous, displaying past him and sometimes thumping him on the back. Our biggest fear is that in his panicked state he will wrench his arm, breaking it. Or a troop member, in an attempt to help free him, will pull on him, again twisting his arm and injuring him.

We quickly get to him, talking to him in a soothing tone. His mouth is open, he is whining. Then he breaks out into another scream. The look in his eyes is one of utter terror, of an animal that's trapped. For a moment, he moves closer to us releasing the tension briefly on his arm, and we quickly push his arm up so he can extract it. He scrambles away in search of his twin brother. This is just like when a little kid sticks his head through the spindles of a bunk bed frame or through the stair banister. Once through, they find it is nearly impossible to extract themselves. That's when Mom yells to grab a stick of butter or the can of Crisco and bring it to her so she can oil up your wayward brother's head.

It breaks our hearts to see one of the gorillas frightened or vulnerable, and after a few shaky laughs we sit back down at the kitchen table to catch our breath. Dianna is already on the phone to maintenance about getting the door fixed. We will mesh it over so little ones can no longer slip their arms through. This incident reminds us of just how vulnerable these animals are in our care. It is not unheard of for keepers to arrive in the morning at other zoos only to find the lifeless body of an infant or juvenile who got wrapped up in rope or a fire hose that had too much play in it, hanging themselves. These are the things that we keepers think about constantly. We look at one another and read each other's minds, "What if that had happened at night?" We don't need to or want to say the words out loud, as if by saying them we might unwittingly unleash the bad luck of it all.

A part of our job as keepers is to give lectures. On a rainy Saturday I drove over to the zoo's education building to give a talk to a group of students from near Kent, Ohio. I think it was group of university-level kids although there were some adolescents in the audience as well. It was late fall so the grounds were empty. There is something so lovely about the quiet of the place on days like this; it's as if you can breathe more easily. As I left the lecture and begin the drive back to the Ape House, I see a young girl who had been at my talk, maybe around twelve or thirteen years old, walking with a woman I presume to be her mother. I have no idea why I stopped, maybe I took pity on them with the rain threatening to begin again, but I pull the golf cart up next to them and ask where they were headed. "To the Ape House," they replied. So I tell them to climb aboard. As I drive over, I inform them that there is very limited viewing for gorillas as no gorillas would be going outside today, but Oscar's building is open and the other building afforded some viewing but only if the gorillas chose to hang out in that particular enclosure—as they were always given the option to be off exhibit.

There is something quite lovely about these two—this mother, daughter team. They are clearly fascinated by gorillas, enthralled really, and I greatly appreciate the mother giving the young girl the space to ask her questions. By the time we arrive at the building, I am already charmed by them, so I invite them to come under the back shed overhang so they can view gorillas. This is an off-exhibit area to the public but still outside the building. It's where we store bales of fresh hay and big heavy metal barrels of soiled hay from our daily cleaning. I excuse myself and go into the kitchen to let Adele know I am back. Del comes out to meet the mother and daughter. We kind of look at each other over their heads and are instantly on the same wavelength. Adele and I head back in the kitchen for a quick discussion. We rarely bring the public into the building unless it is a tour requested from the powers that be, and this girl is a little young, but we keepers have discretion on these types of decisions. Adele has a good feeling about these two as well, so I go outside and formally invite them into the building. Once we are standing in the kitchen, they immediately see Colo at the back door of her enclosure and they can barely contain themselves as they try to concentrate on what I am currently explaining to them: do not look them in the eyes, do not talk to them, do not go near the enclosures, do not react if they throw something, do be quiet and respectful.

The four of us head around to the front bench. It's not that we keepers ever really forget what an honor it is to work with gorillas, but anyone can get a little jaded at times with the workload, the physical and emotionally taxing work, the inevitable frustrations in any workplace. But we see the gorillas through the eyes of these two neophytes, who are both awed and amazed to be in such close proximity to gorillas. We hear it in their hushed and respectful tones and we too once again feel the wonderment.

38

STORYTELLING

In the spring of 1988 in the midst of seismic husbandry changes and my first quarter back at The Ohio State University, filmmaker Allison Argo contacted us. Allison's grandmother lived in Birmingham, Alabama, and as such she often spent time there. During one of her visits, Allison met Pongi's keeper, Randy Reid, and their frequent conversations impressed and intrigued her. Allison wanted to film a documentary about captive gorillas using Randy and Pongi's relationship as one of the seven interwoven stories. But the lynchpin would be an adult male named Ivan and the focus would consistently loop back to his particular story, his life in captivity. Ivan was famous for being the "Shopping Center Gorilla" who was on display at a strip mall in Tacoma, Washington. Ivan's life of solitude, living in a concrete room devoid of social stimulus or companions, was the centerpiece. The six other gorilla stories radiated out from Ivan's story, showing the progress we had made while exploring the intricate relationships we had with these animals.

Randy Reid had taken care of Birmingham's two resident gorillas for years before they were shipped off to other zoos, and he had not seen Pongi (or Susie as she was called at Birmingham) since her move to Columbus several years before. Much had changed for Pongi. She was a mother now, raising her five-month-old son, Colbi, within her small troop. Allison thought that a reunion of Randy and Pongi would be an interesting take, and we wholeheartedly agreed.

We spent the evening at Dianna's house sitting in the living room, munching on snacks and drinking champagne. We listened to Allison as she explained her documentary concept, then we brainstormed, throwing out our thoughts. I suggested she contact both Howletts and Apenheul to do one or two segments, as they were way ahead of North American zoos at the time in terms of philosophy and husbandry practices. My hope was that by highlighting these facilities, US zoos would be exposed to another way of looking at gorilla husbandry, facilitating much-needed changes within our zoo community here in the States.

Six months later in September 1988, the film crew arrived with the delightful and gentlemanly Randy Reid. As we prepped the outdoor habitat, Randy kept a low profile and was not allowed anywhere near Pongi in order to preserve the full element of surprise when she finally saw him. Nancy was there to help film and also served as a backup to Allison's then-husband cinematographer Bob Collins. And Julie Estadt from our marketing department was there to take still photos of the filming process and reunion.

Pongi has a distinctly formidable look to her. She is a small female, relatively speaking, but there is a self-contained aloofness to her when interacting with keepers. In essence, she was a gorilla and as such she conveyed the sense that we were there to serve her. To feed her, clean her enclosures, and then remove ourselves from her gorilla life. She always reminded me of a very strong human matriarch who would brook no nonsense from other family members. When she spoke, everyone listened. It wasn't that she wasn't capable of being goofy or funny on occasion, it's just that this side of her wasn't usually directed at us. Pongi was not one to normally solicit play with a keeper; she was not one for interactions with the keeper staff in general. She did not need us, and I respected her all the more for it. She was a gorilla, through and through.

On the day of the shoot, we planned the reunion shot around transferring Pongi via a short underground tunnel over to the habitat from the north yard. Because it was a blind spot, she wouldn't be able see Randy initially. None of us knew what her reaction might be, but we knew we had only one opportunity to capture it. My role was to operate the doors to transfer her, and Dianna would give me the go-ahead once the film crew and Randy were in place. Because I was giving Pongi access to the habitat yard, I had a full view of her

FIGURE 38.1. Randy Reid with Pongi and Colbi, during filming
of *The Urban Gorilla* documentary

before and after shifting, allowing me to observe her body language
up close. I opened the door and watched as I heard Randy call her
name, "Susie," sounding like "Suseh" in his beautiful soft Alabama
accent. Pongi's body language instantly changed; her hair coat stood
up (something that happens in primates when excited or scared,
called *piloerection*), and then she ran at breakneck speed in the direction of Randy's voice, all the while making continual low rumbling vocalizations. By the time I got around to where Randy and
Pongi were, Randy had his hands through the mesh and was touching and talking to her. The whole time Pongi had eleven-month-old
son Colbi on her back, the usual form of transportation for an infant
that age. She seemed to us to be presenting Colbi to Randy, who was
freely allowed to touch him. She had rarely if ever allowed us to touch
Colbi. There was not one dry eye amongst us, the film crew, the keepers, the photographers, or Randy.

Here are Randy's impressions in his own words:

After lunch everyone was in place with their equipment and
the big moment finally arrived. Dianna and her crew would

try to shift Susie and her baby, Colbi, into the habitat alone, but Oscar came first. Standing at the rear of the habitat, I was about twenty yards away as Susie came through facing in the opposite direction. I called. Upon hearing her name, she spun around and with no hesitation ran to where I was standing. What a great feeling!

I wish I could describe that moment. Without doubt, this was the high point of 25 years as a keeper. We vocalized to one another, and she allowed the baby to sniff and touch my hands. After the lengthy separation, two old friends met again and a dream for the human came true. She did remember!

It was so moving and touching that when the documentary aired a year and half later, it illuminated in its utter simplicity just how remarkable, complex, and long lasting our relationships with gorillas are. In all my public lectures I use that particular clip to illustrate the depth of some friendships that exist between gorillas and people.

Sometime in 1994, the Columbus Zoo was asked to participate in a summit in Seattle concerning Ivan. Allison Argo's documentary had brought Ivan's story to the fore and many in the animal rights and zoo communities rallied around the idea of getting Ivan into a different setting. And because the B & I shopping center in Tacoma where Ivan had resided since 1962 was filing for bankruptcy, the timing seemed right. I accompanied our veterinarian, Dr. Lynn Kramer, to Washington, where we joined the summit organizers—Ron Kagan, director of the Detroit Zoo, Dr. Terry Maple of Zoo Atlanta, and Mitchell Fox of the animal rights organization PAWS (Progressive Animal Welfare Society), which was truly the earliest voice for Ivan and the driving force for this meeting. The hope was that Ivan's owner could be persuaded to relinquish ownership of him to a North American zoo and that Ivan could be placed at a facility that would allow him access to other gorillas and an outdoor setting. This meeting gave us the opportunity to meet Ivan, assess his situation, and discuss possible options of where he might live.

Initially, Columbus Zoo was in the running with Zoo Atlanta to possibly house him as by this point both institutions, especially Columbus, had extensive experience and success with gorilla introductions. But prior to leaving Columbus, our staff met to discuss the situation and we all agreed that Columbus simply was not in a position to take on another silverback. We already had four adult males and that was proving problematic. But Dr. Kramer and I went anyway as I could add my thoughts and suggestions from a keeper's perspective on introductions, on so-called nonsocial gorillas, and on the adaptability of gorillas when placed in the right setting. Eventually Ivan was sent to Zoo Atlanta where he lived out his life in the company of a number of females, one of whom he bred with although she never conceived. But it should be noted that while he got on well with his females, he still directed much of his attention toward keepers rather than gorillas throughout the remainder of his life.

Telling stories is as old as humanity. Sitting around a warm and comforting fire at night was as much about security and safety as it was storytelling, but surely our ancestors' campfires were their version of our kitchen table, gathering together at the end of each day to delight one another with experiences—to amuse, entertain, and teach.

Within the next ten years, more storytellers would approach us to ask if they could help tell additional stories of these extraordinary animals we cared for. A producer from New York City, and a local TV cameraman, Dan Friedman, began filming another documentary focused solely on the Columbus Zoo's gorilla program. *Baby Gorillas: A Gorilla Family Portrait* began filming in 1991. It explored the brutality of gorillas captured from the wild to be taken to zoos, zoos ill-equipped to care for them properly, and it included extensive interviews with our staff as to why we wanted to see change. The documentary also showcased mother rearing and our surrogate program. Because we knew Dan, who had worked extensively with Jack Hanna on his show and in some cases with us, and because he did his homework when interviewing us, asking all the right questions, duly noting and taking into account our thoughts and passions on the subject, his script was totally on message. It painted a complete and true picture of daily life in the Ape House. And because Nancy had been documenting the progress of our husbandry program over the years, the production company had a treasure trove of film footage from which to mine all the gorilla milestones. I remember being given a master copy before final edits were made and I took it home to watch. I was moved from start to finish with the respect afforded the gorillas by Dan and the producer. They had nailed it.

39

SOMETHING BIGGER THAN US

Charlene and I are once again at work on another coauthored paper. In June 1989 Emory University will be hosting the Fertility in Great Apes conference. Our presentation in Atlanta will not be from a hard-core scientific data perspective but rather a fly-by-the-seat-of-your-pants approach to possibly solving the infertility issue that is presumed to plague the North American gorilla population. We are focusing on the gorilla's innate social skills to solve their own problems, when they are placed in situations that allow them to be gorillas. We will be presenting a paper on the surrogate program, and as a coauthor, I will be attending and Charlene, as usual, will present the slide presentation.

In preparation for the conference, we also designed and printed flyers announcing that the Columbus Zoo would be hosting the first-ever Gorilla Workshop the following year, in June 1990. Our trip to Atlanta would give us the opportunity to talk to people and to get them jazzed about the workshop.

I did not have a computer at my house so I spent almost every evening after my night classes at Charlene and Bob Jendry's house typing madly away on their computer late into the night. I wrote letters to potential speakers, drafted workshop updates, and worked on the schedule of events. Adele, Charlene, Dianna, and I sat down at the kitchen table and as a staff we scoped out who we would like to see attend the workshop, how to promote it, who we could get as sponsors, who we needed to contact, and who was to do what. We'd make a new checklist after discarding the old one, listing what's been done thus far and what still needed to get done. Adele was our Australian contact, as she had studied there, and she was instrumental in enacting a keeper exchange program between the Melbourne and Columbus zoos. Charlene was our contact for the Mountain Gorilla Veterinary Project (MGVP) and was able to get Ruth Keesling of the Morris Animal Foundation, which oversaw the MGVP, to be one of our keynote speakers. Other speakers included field researchers and academicians Drs. Kelly Stewart and Sandy Harcourt of the University of California-Davis, Craig Sholley of the Mountain Gorilla Project, and primatologist Ian Redmond, who had worked closely with Dian Fossey. The keynote speaker list was a little top-heavy on mountain gorilla studies, but at that time there were few established western lowland gorilla field studies.

One of my goals was to get keepers Pete Halliday from Howletts, Frans Keizer from Apenheul, and Richard Johnstone-Scott from Durrell Conservation Trust on Jersey Island to come and present lectures. I strongly believed in their respective husbandry programs and felt that North American zoos had much to learn from their common-sense approach to captive management. In the end, we were able to get Frans from Apenheul and Richard from Jersey. And even though Jack Hanna had sent a letter directly to John Aspinall, we were unable to get a Howletts keeper, much to my disappointment. The good thing was Richard was a former keeper at Howletts, so he was able to lend his perspective on both Howletts and Jersey Island.

Luckily, we had a fabulous group of seasoned conference hosts in our docent organization. They had recently hosted their national docent association conference in Columbus, and their practical guidance and creative input on how to host a successful conference was invaluable. They taught us that being the host was as much about the atmosphere as the caliber of talks. We were also committed to making both the registration fee and hotel rates extremely affordable. This formula is one I have carried with me for all types of conferences I have organized since, and it has proven to work in every case. Make the conference affordable, but most importantly, make attendees feel like welcome guests from the moment they register to the last day of the conference. It's analogous to hosting a dinner party: every single detail matters.

A conference is successful when it builds a community during the event and sustains it afterward. Several key components are essential. Keep the conference small, no more than three hundred people. Never have concurrent sessions so that everyone is always together. And provide as many meals as is possible. Standing in a buffet lunch line is a great place for people to introduce themselves to one another, and networking naturally results. I would venture a guess that many of my gorilla keeper colleagues met and developed long-lasting relationships with others while standing in food lines at Gorilla Workshops over the past twenty-five years.

As registrations began coming in, we bought a world map and placed it in the Ape House using multicolored stickpins to track how many countries were represented. As each new abstract was submitted and more and more people got wind of what we were doing, the anticipation built. It was exciting to take a break after the morning cleaning and feeding to pick up our mail at the zoo's main office and see who else had registered. With the additional work and planning of the conference, we were lucky that we had a fantastic summer keeper staff in Liz Garland and Laura McMahon. They were a huge support during the organizing process and freed us up to get our conference preparation work done.

On the morning of June 22, 1990, I was scheduled to give the welcoming speech at the workshop. I got up early as I was unable to sleep, sick with nerves at the prospect of speaking in front of two-hundred-plus colleagues, and went to swim laps in the hotel pool, something that usually calms me. But no luck this time so I decided to head out to the zoo.

Before any of the other staff had arrived, I let myself into the Ape House at 6 a.m. For a moment I stood still, closing my eyes as the sounds and smell washed over me. Then I went and sat in front of Bongo and Fossey's enclosure. I'm sure Bongo was more than a little baffled by my early morning intrusion. He sat patiently, a bit puzzled perhaps, while I just hung out. Being in his presence, I was reminded why we were hosting this conference. Once again he was my direction when I lost my way. All of it—the workshop, the *Gorilla Gazette* newsletter, and our surrogate program—was for him. And just as someone told Dian Fossey years before that she was the keeper of the stories and as such must share them, I saw with clarity that this wasn't about me and my nerves and insecurities. This was about Bongo and so many others like him.

The workshop was a chance for the people who cared for captive gorillas to give voice to their charges. We could collectively change husbandry and make improvements. We could show management that mother and father rearing should be the norm and that bedding was not an option for gorillas but a must. We could show that complex and integrated exhibit design was a necessity. We could show that these animals were naturally social when given the simple opportunity to interact with conspecifics. The Gorilla Workshop was about changing the captive gorilla world as we knew it.

40

FRIENDS

In September 1986 something unexpected happened at the Jersey Zoo: a young boy fell into the gorilla enclosure. Knocked unconscious by the fall, the boy lay sprawled at an awkward angle in the cement moat as Jersey's twenty-five-year-old silverback went over to see him, accompanied by an inquisitive juvenile. The entire episode was documented from start to finish through videotape taken by the public.

Jambo sat on the raised cement edge of his grassy yard and reached down into the moat below to gently touch the exposed skin of the little boy's back and then sniffed his fingers. The crowd reacted in restrained horror as they saw him reach for the boy, but then seemed to recognize the inherent gentleness in the gesture and quieted down. Curious and more than a little spooked by this unusual and unexpected "thing" in their exhibit, one of Jambo's females also walked over with an older infant clinging to her back. Drawn by the boy, she worriedly peered around Jambo to get a better look. Jambo placed his body subtly in between his female and the injured boy and then seemed to actively shoo the female away.

Jambo continued to sit several feet away from the boy, blocking any troop members from him. With his back to the crowd, Jambo seemed to peer over at the boy with a curious tilt to his head as if thinking, "What's this?" The boy started to wake up, obviously disoriented and confused, and initially began to whimper in pain. But when the boy started to cry and the crowd began to yell and shush at him to stay quiet, Jambo's demeanor changed and he began to look nervous. Gorillas in general don't want trouble. The boy was now fully awake and started to cry in earnest. It was then that Jambo chose to make his exit, quickly leaving the scene, glancing uneasily behind him several times as he hurried away. By this point, one of Jambo's keepers, Andy Wood, and another keeper from the bird department, Gary Clark, entered the enclosure. Their presence allowed the emergency worker to reach the injured boy and safely remove him. The boy fully recovered from a concussion and several broken bones and, along with Jambo, became forever famous.

Back in the 1980s, many of the pioneering husbandry programs could be attributed to several European programs with their inventive directors and keepers. Head gorilla keeper at Jersey Zoo, Richard Johnstone-Scott, was one of these keepers and was an eminent and esteemed figure to many of us in the zoo world. He had extensive experience as head keeper at Howletts, prior to going to the Durrell Conservation Trust (Jersey Zoo), where he was again instrumental in shaping their gorilla program—allowing keepers to enter the gorilla enclosure when necessary. But while at Howletts, Richard helped to raise our surrogate dad Mumbah when he arrived from the wild as a young frightened four-year-old gorilla. Richard eased Mumbah's transition to captivity, and they had a long-established friendship together.

It is the day before the Gorilla Workshop. After almost two years of planning, organizing, and finalizing, all that we had been working toward is finally here. This four-day international conference, bringing together gorilla keepers and gorilla researchers for the first time, was a direct offshoot of our publication *Gorilla Gazette,* which we started in 1987. The idea of *Gazette* came to me one day when realizing that as Columbus Zoo keepers, our voices were readily listened to but we were the exception rather than the rule. When I first brought the idea to head keeper Dianna and then to fellow keeper Charlene, we all immediately saw the possibilities and enormous influence a publication for and by gorilla keepers could have. If keepers at other zoos were not able to have a voice at their own institutions, then we would provide the format for them—through a publication. Less than two years later I spoke to Dianna and then approached Jack about the idea of possibly hosting a conference. I remember it clear as day, catching up with Jack as he made his daily rounds of the zoo— walking at breakneck speed trying to keep up with him as he juggled any number of issues that needed addressing. I explained what we wanted to do, that *Gorilla Gazette* had taught us that there was a real need to communicate about gorilla husbandry issues. Without missing a beat, he looked at me and said, "Get something to me in writing so I can give it to the board." And that was it, easy peasy, the first Gorilla Workshop was born.

Three years after the inaugural issue of *Gorilla Gazette,* our first Gorilla Workshop is about to get underway. I have just come back from one studio interview with a local television station and will soon be called away for another one here on zoo grounds but wanted to stop by the Ape House for a moment to see how things are going.

Our guests have begun arriving for behind-the-scenes tours before the start of the conference tomorrow. It is incredibly satisfying to finally put names to faces. Many of those attending have contributed to *Gorilla Gazette* over the past few years, so we know their reputations and their vast and varied experience with gorillas. Violet Sunde from Woodland Park Zoo is one of the first to arrive. I would recognize her distinctive, slightly gravelly voice anywhere, having had numerous conversations with her over the phone. She is a tiny, slim woman with a quiet calm but fierce resolve. She is committed to doing right by captive gorillas. There is a definite buzz in the building as more colleagues begin to arrive. The expectation of meeting fellow keepers known only through correspondence until now is palpable. And underneath it all, above it all, intertwined in it all is the feeling that this is the place to be—that something extraordinary is about to happen.

Richard Johnstone-Scott arrives. He is wiry and lean, his face tan, and his accent unmistakable. We give him a quick tour of the building, but he is anxious to see his old friend Mumbah. When Richard walks out into the keeper area, Mumbah is leaning up against the vertical support beam of the habitat, his usual spot. Richard calls his name. Mumbah's reaction was not surprising, but his quick response time was. Mumbah, who interacts with us, more out of necessity than want it seems, immediately ran over. Then he and Richard sat on either side of the mesh, head to head, oblivious to all else around them, looking like a couple of cronies exchanging long-held secrets as they settle in for a much-anticipated visit.

Richard described the reunion in a 1990 *Gorilla Gazette* article:

Seeing him again that Sunday afternoon at the Columbus Zoo was very special for me. He looked so well and, despite the sudden en masse appearance of us "Workshoppers" came straight over in magnificent style and remained relaxed and in close attendance for most of the visit. Whilst keeping an eye on the sprightly Colo, a famous individual I have always wanted to meet, I mumbled a few hopefully familiar sounds to Mumbah. We comfortably held eye to eye contact, there being no threat intended or taken; then he gave a priceless albeit barely audible grumble! He was so near, yet so far, and I found myself for purely selfish reasons wanting to desperately put an arm around him, and ruffle his crest, but I refrained. I know from experience that it can often be "difficult" showing people around behind the scenes, homo sapiens being such unpredictable creatures, and I hasten to thank the staff for being so tolerant and understanding. Later in the day I returned to watch him from amongst the crowds, he was interacting playfully with a youngster, their mouths agape with pleasure—a heart-warming sight. It is gratifying to know that Mumbah is now enjoying life to the full with his adopted family and such pleasant surroundings. He is obviously very contented.

41

MOSUBA'S ROAD TRIP

The back of the zoo van reeks to high heaven. Mosuba, a soon-to-be seven-year-old gorilla, is sitting quietly in his small traveling crate, his arms crossed over his chest, looking somewhat bewildered. When stressed, gorillas will literally shit themselves silly. Mosuba, anesthetized earlier that morning for his final physical, had been separated from his twin brother, his constant companion since birth, and was now being transported across country. He was, to put it mildly, anxious.

For some reason, our veterinarian had vetoed the need for hay, so Mosuba had been placed in an empty transport crate. The hay would have absorbed some of the feces and accompanying odor as well as provided a more comfortable setting for his fourteen-hour road trip. Think of wrapping yourself in a blanket when frightened or worried; hay is the same, providing comfort and warmth. It was a baffling decision on the part of our vet, and although we had argued for bedding and lost, it was Mosuba who was suffering because of it. Add to that it was late summer, and the van's air conditioning was hit or miss. The attendant results were pretty powerful. I spritzed baby powder every once in a while to cut the stink.

Because of the unusual circumstances of the twins' births and our mutual commitment to keeping them together, both the Columbus and Henry Doorly zoos agreed to "share" the twins. They had spent all their lives going back and forth between Columbus and Omaha twice a year. We had named the first twin, Mosuba, telling folks at Omaha that Mosuba meant "travel between two villages" in some obscure African dialect, but in reality he was named for the three volunteer keepers who helped to raise him: Mo for Molly, Su for Sue and Ba for Barb.

After being nursery reared for three and a half years, the twins were integrated into a troop of gorillas, led by our ever-patient silverback, Mumbah. At the time, they were the youngest gorillas we had introduced back to adults, and they served as a bridge to building our fledgling surrogate gorilla program. The knowledge we gained from their introduction led to us being able to integrate even younger infants into the troop in subsequent years.

The twins had become such an integral part of our ever-evolving growing troop. We felt that removing the boys every five to six months was proving to be too disruptive to the group as a whole and to the twins themselves. We started to push to have both boys stay at Columbus on a permanent basis. We wanted to keep the twins together for another reason, too. The birth ratio of captive gorillas is one to one, half males, half females, which meant that in the not too distant future there would be far too many males and far too few females to form normal gorilla groups. In the wild, a male gorilla will have several females and their offspring within any given troop. We felt that keeping the twins together might prove to be a model of co-leadership that would help alleviate the issue of how to house excess males. The twins' behavior would dictate whether that would work down the road, but we definitely thought it was worth a try. Omaha, however, differed and wanted their twin back. So that summer, along with two other keepers, we made the long, painful drive to Nebraska.

Somewhere in Iowa, we had had enough of the stink-mobile and decided to find some hay. We stopped at a Holiday Inn by the interstate. I went in and asked at the front desk if there were any farmers nearby that might be willing to sell us a bale of hay. The hotel manager informed me that they were having their local Rotary Club meeting just around the corner in another meeting room and that many of the Rotarians were farmers. I walked in the meeting room just as the Rotary president was finishing his announcements. When he was done, I walked up, a bit hesitant knowing I smelled to high heaven of gorilla and explained our situation: "We have this young gorilla in our van, he is pooping himself silly, we are from the Columbus Zoo, we desperately need hay. . . ." He immediately left the meeting, asking us to follow him. He was our farmer.

Several miles down a country road, we pulled into a gravel drive, with a neat white farmhouse on the left and an equally neat white barn to the right. The farmer went into the barn and came out a few minutes later with half a bale of hay, then proceeded to go inside his house. We opened up the back doors to the van, leaving them open to air it out, and put the hay in with Mosuba, who promptly started fluffing it up, making a nest.

Moments later the farmer and his family came walking toward us. He had three children, two boys and a girl, all a little tentative. There was a look of curiosity mixed with—not quite wonder—but something analogous to it on the younger girl and her older brother's face. But the middle boy was different; his gait was a bit stilted, the look on his face more inward than outward. He hung back a little until his father placed a reassuring hand on his shoulder, urging the boy forward.

I have often thought of that day over the years and wonder if the day a gorilla came to their farm to get some hay is ingrained in the fabric of this family's history, a part of their mythical lore, to be forever passed from one generation to the next. The day when this kind farmer's middle son and a young gorilla named Mosuba calmly looked at one another, taking each other's measure in thoughtful appreciation, as if in recognition.

42

PEOPLE I HAVE MET

Recently I saw something a friend had written in her notebook that she posted on Instagram saying, "keep good company" in her thoughtful cursive handwriting. It was a simple but powerful statement.

Being a keeper brought more than just the presence of the gorillas into my life; it opened me up to a world of like-minded people, all of whom had a deep love and appreciation for great apes. Fellow primatologist and friend Ann Pierce recommended a book by Tim Cahill, called *Jaguars Ripped My Flesh*. I was hooked from the moment I picked it up and then proceeded to read anything Tim had ever written. My father, too, became a big fan of Tim's writing, especially of the travel adventures Tim shared with National Geographic photographer Michael "Nick" Nichols, another person Ann introduced me to. Ann got to know both Tim and Nick when they visited Karisoke Research Camp in Rwanda in the early 1980s, where she was studying the fringe groups of mountain gorillas. As a testament to Tim's storytelling skills, I give away his books like they are candy. For me they are on par with books like *Boy's Life* or *A Story Like the Wind*, books that suck you in from the opening sentence, not letting you go until the very last word.

One of the best stories I ever read about gorillas was by Tim. It was dramatically titled *Love and Death in Gorilla Country* and appeared in another collection of his travel essays, *A Wolverine Is Eating*

My Leg. Don't let the title put you off; it's tongue in cheek, a nod to the sensationalized adventure writings of the 1950s. And don't let the humor distract you. Tim is a beautiful writer. He can capture the absurd in his funny, irreverent way while two sentences later he'll make you cry with his sensitive, perceptive take on life. Here is an excerpt from his story—where he has completely captured that feeling that one gets in the presence of gorillas.

> The gorillas let you know when you have overstayed your welcome. They let you know with frowns, locomotive coughs, the beginnings of displays. Even so, there was a sense of enormous privilege just sitting with them; privilege felt in their acceptance, no matter its duration. Once, on the lower slopes of Visoke, one of the volcanoes, a silverback named Ndume woke up from his nap, saw me, and ambled close to the place where I lay.
>
> Ponderously, he sat down, yawned, stretched, then reached down to my knee, where he took the material of my red rain pants between his thumb and forefinger and rolled it back and forth like a knowledgeable garment buyer. Ndume cocked his head to his side.
>
> His eyes were soft, golden brown, and he wore that familiar slightly puzzled expression. "Gore-Tex," I wanted to say. Instead, I grunted twice. Ndume returned the DBV (double belch vocalization). There was some genuine interspecies communication going on, and it felt like a fantasy, like one of those strange wondrous dreams in which you can talk to the animals, to all creation, and creation itself responds with approval. There was no science involved in this encounter. It was all emotion. I felt an unspecified glow, something within that was very much like love, and it came to me then that for the past fourteen years Dian Fossey had literally lived in that glow.

Visitors to the Ape House were sometimes planned and sometimes not. One never knew who might show up and almost, without exception, it was a pleasant happenstance. It was early autumn 1988 when the front office called to say that two gentlemen were here to see the gorillas—Dr. Tom Begg, the veterinarian from Howletts, Mumbah's home zoo; and another vet, Dr. Bob Cooper. Both Bob and Tom had just attended the World Conference on Breeding Endangered Species in Captivity hosted by the Cincinnati Zoo.

I gave them a tour of the Ape House, explaining our philosophy and husbandry. Tom was especially interested in seeing Mumbah and was pleased to see his troop growing under Mumbah's gentle tutelage. But it was Bongo that seemed to have the greatest effect on him. As we sat in front of Bongo's indoor area watching Bongo interact with his young son, and I explained their circumstances—Bridgette's untimely death, how Bongo was raising his son Fossey alone—I will never forget Tom's quiet response, "Oh yes, I've heard of this. It's a lovely story."

They finished the tour—a nice break to my day—and then I got back to work. Later in the afternoon while working outside, I looked up only to see Bob Cooper standing there, saying he wanted to stay another day to tour the zoo in order to get a better feel for it. I became so engrossed in speaking with him that the gorillas started to get annoyed at the delay of bringing them in, so I told Bob, "Let me get them in, fed, and settled, then I'll have you come in the building."

Sometimes you meet people that you just click with on an intellectual level, and that was Bob. After sitting for quite a while discussing our gorillas and his experiences in Gabon, I was reluctant to end our conversation, so I told him he was welcome to come have dinner with me on my way home and that morphed into a three-day stay, filled with great conversations about his life, his recent work in Gabon, his wife, Sian, and their two children. Bob was scoping out job opportunities here in the United States while Sian was in her native Wales with the kids, visiting relatives. It's one of those great gifts that come unexpectedly, courtesy of the communal pull of gorillas.

I recently found an entire file folder of letters and cards from late 1988 through 1991. In the pile were loads of letters from Bob in his distinctive print with its strangely elegant cursive look to it. One letter in particular gave me a clue as to why perhaps Bob and I had such a strong connection. He mentioned that he had applied for a job at a zoo out West and that one of the people interviewing him had an issue with him. Bob had apparently ruffled a few feathers at the captive breeding symposium when he voiced concerns about a particular great ape species. Bob was told, "It would be difficult to work with zoo people with whom one had been so overtly critical/caustic." Ahhh, the wrath of zoos back then when questioned or critiqued.

That may have been the connection—our passion to voice what others may be thinking but were reluctant to say. Bob also mentioned that this particular fellow who castigated him was without humor but ambitious, which was to me a deadly combination if there ever was one. I have watched over the years the obvious ambition of a few colleagues in the zoo world; their machinations were as transparent and detectable as watching the social dynamics play out in a gorilla troop. I have nothing against being ambitious, but I do recognize its limitations. Ambition can temper your response to difficult issues. It makes one reluctant to speak out—for fear of being labeled "troublesome." It makes one hesitate because of considerations to one's own career, which might be all well and good in the business world but when dealing with animals, not so much. With captive animals, we are their voices—we speak for them by placing ourselves in their position—we are obligated to speak up and out for them.

Other visitors that delighted us were fellow gorilla keepers. Some came to pick up animals, always difficult for us. Others came to drop off their gorillas, always difficult for them. As the Columbus Zoo was one of only a handful of officially sanctioned primate quarantine facilities in the United States, we were often asked to house primates when they entered the country. Quarantine rules stated that an incoming primate from another country must be placed under strict quarantine for a period of thirty days before being cleared to mix with the population in US zoos. In 1994 Toronto Zoo gorilla keeper Vanessa Phelan brought their two young male gorillas, Patrick and Jabari, to us for quarantine.

We spent many hours at my house as we talked about our philosophy and husbandry practices. Vanessa, a tiny woman with a lovely Aussie accent, delighted us with her wry and incredibly funny take on life. It was an absolute joy to be around her as we shared our thoughts. In the 2011 *Gorilla Gazette* issue, Vanessa wrote of that time, "We took back with us to Toronto the determination to do as they did in Columbus. It was time the gorillas lived together day and night. 20 years of being separated at night from each other had to stop. It took a year to accomplish. I think it is important to stress that if it had not been for you and Charlene and what you shared with me back in 1994, Toronto's gorilla family would have faced a very different future. Eventually things would have improved, but those few days with you two changed their lives completely."

43

MOLLY

Molly loves to comb her hair. I wore a plastic headband to work today, the kind with the short stiff bristles. Quick as you like, she grabbed it off my head, broke it in half, and is currently running it through the hair on the top of her head, giving herself a coif, it seems.

In 1988, we received thirteen-year-old Molly from the Kansas City Zoo. She is perhaps one of the most gorgeous female gorillas I have ever seen. Her looks came from her handsome father, Trib, who was housed at the San Diego Zoo. Unbeknownst to us, she had skills beyond her good looks.

Molly was moved into the north (Oscar's) building to be introduced to his troop. This building has a long rectangular room with a public viewing window running the full length of the room. The exhibit was a dull gray, coated in sprayed gunnite—the usual look back then to make it look like simulated rockwork. Why we thought a room full of simulated rocks for gorillas was an appropriate approximation of their life in the wild is beyond me, but there you go. Gunnite has a rough texture giving it a grip. The water bowl located on the far left of the exhibit is nestled high in the corner and is fed by a slightly raised waterfall, creating a cascading outcropping.

There was an attic of sorts in Oscar's building that the keepers had access to from the public aisle through a locked door and then up a ladder. The attic allowed us to look down on and observe the gorillas by way of simply sliding back a rather flimsy particleboard panel. There was no lock on the panel that I recall, nothing secure about it except that it was quite high up. That in and of itself was deemed sufficient to keeping gorillas in their exhibit space.

Bill Cupps and Diana Frisch taught me to play the card game peanuts on winter afternoons in the attic of that building. It's a game in which each player has his own deck of cards. In essence, you are playing solitaire but with a communal pile as well. It's collective solitaire on steroids. It is an incredibly fast-paced and intense game with the goal of getting rid of all the cards on your own peanuts pile first. Many years later, I taught my nieces and nephews how to play when they were very young. It is still one of our favorite family card games, although the kids like to remind me that I used to scare the bejesus out of them with my ferocity while playing.

But in the keeper attic, we had never figured on a gorilla named Molly. One day, Molly stood on the top of the raised corner water bowl, pushed against the opposing walls using her hands and feet as pressure points, and then climbed to the top of the exhibit like a seasoned rock climber. When we opened the attic observation window, it was quite a surprise and somewhat unnerving to see Molly directly across from us, staring at us from fifteen feet or so away. Luckily, she had not yet figured out how to scale the much smoother opposite corner wall, which was well within reach of the attic's window opening. We wisely decided we needed to move her back over to the main Ape House where each indoor enclosure was meshed over.

Molly's life with us continued to be eventful. Once when doing annual physical exams, a visiting veterinarian accidently collapsed Molly's lung when intubating her. Let me say this, anaesthetizing a gorilla is always precarious at best, and we would lose some animals, as have other zoos, while knocking them down for "routine" annual physical exams. I, for one, do not think it's worth the risk and find the use of the word "routine" inappropriate and misleading.

After her exam, Molly was brought back down to the Ape House and placed in her enclosure, but it was evident to me that she was having difficulty breathing from the get-go. I called up to the vets, who were now back at the hospital working on another gorilla, and explained my concerns. Shortly thereafter, I called a second time, and this time I think they got the message. She stopped breathing just as they arrived and, fortunately, they resuscitated her. As they were taking her back to the hospital, she crashed again. They worked on her and she came back once more.

It was nighttime before they finally got her stabilized and brought her back to the Ape House. She was placed in an enclosure, and initially, I sat outside softly talking to her as she slowly, slowly came to. But she kept reaching for my fingers through the mesh, so I went in with her. The vets were there as well, and we decided that my being in close proximity would be a comforting presence to her. I lay down next to Molly, as she reached out her hand and I held it while talking to her in a quiet soothing voice. Do I think it was me she wanted? Not really. I think any of the keepers would have sufficed—I just happened to be the one on duty. But I will say this, it took me back to that time as a kid when my cat wanted me there while she gave birth. I felt that same sense of wonder and privilege.

After that, we made it standard practice that there would be an anesthesiologist at every gorilla knockdown. We already used human doctors during exams—a cardiologist, dentist, neurologist, neonatologist, obstetrician—so it was a logical addition. Gorillas are like us internally, very little difference, so it seemed that the most prudent thing to do was to hedge our bets whenever anaesthetizing gorillas in the future by using a human doctor.

Molly was eventually integrated into Sunshine's troop, and in November 1992, she gave birth. As with any first-time mom, we had no pattern of her maternal skills and behaviors, so we needed to rely on very concise observations. At the time of the birth, I had changed jobs, taking the head keeper position in the Australia/Asia department immediately after graduating from The Ohio State University the previous March. But I was called back in to assist with observations.

The dance between a newborn gorilla and its mother is akin to an older couple (my grandparents who have been married for years come to mind) when they come together on the dance floor. They intuitively know one another's every move, shifting smoothly from one delicate dance step to the next, the husband's hand applying subtle pressure to his wife's back, a gentle and unseen turn to the left or right, each responding in kind.

Immediately after birth, a mother will gently place the infant on her chest, cupping its tiny rear in her hand, giving support to the somewhat astonished newborn. As importantly, and for this dance to flow, the infant must have a strong grip, grasping tenaciously with both hands and feet to its mother's hair coat. Rooting begins instinctively and fairly quickly, the infant's head swiveling back and forth, back and forth, until it latches on the nipple to begin nursing. It is the best of moms who adjust a shoulder, an angle of her body, to position the baby in order to allow easier access to the breast.

Infants, when satiated with their mother's milk, will literally pass out and rouse only when their belly needs more nourishment. While the infant sleeps, the mother may sit cross-legged with the baby safely nestled in the crook between her leg and crotch. It is at these times that conscientious mothers gaze tenderly at their infant, inspecting ears, cleaning goo out of its eyes, lifting a hand or foot to examine a teeny tiny finger or toe, giving their baby a thorough going-over, a full-on-maintenance check.

In the middle of the night, I have been privileged to observe a mother and her newborn in the quiet of a dimly lit building. The mother looks down at her clinging son and pulls his hands and then his feet away from her own body to carefully examine each appendage. When the baby squeaks in protest, the mom vocalizes a soothing rumble and quickly gathers the infant back against her body in a warm embrace—as if in guilty apology, as if she simply cannot help herself, her delight in her infant is just too obvious.

As keepers, it is our job to observe, to note when interactions occur and the length and time of nursings. We note when an infant passes its first tar-like black stool called meconium and when the stool then changes to a milk stool (gold and glue-like), indicating that, yes, indeed the infant is getting milk. We also chart the number of urinations and the rate of respirations. We take note about how the mother responds. Does she respond when the infant vocalizes, or does she ignore the protestations of her offspring? In noting and compiling all of these behaviors, we inevitably see patterns, and these can help us ascertain whether the infant is thriving or not.

The problem for the keeper staff (and ultimately Molly and her infant) was that we were blocked from actually seeing the infant nurse. Molly consistently turned away from the observer when nursings were about to occur. We were unable to see the infant latch on to her nipple to confirm the typical Dizzy Gillespie-like in/out blowing of the infant's cheek pouches when the infant suckled. We had no visual confirmations, but there were subtle behavior patterns that indicated that nursing was taking place. Every one to one and a half hours, the infant would wake, vocalize, and become fidgety. Molly responded by shifting the infant to a good position to nurse and then turned her back to us. No more fussy sounds from the infant, and when Molly did turn around, the infant would be fast asleep, head back, mouth open quite close to the nipple. These were all indicators that the infant was replete with milk, but we could not prove it. She was nursing the infant in theory, but we needed confirmation, or the infant might be pulled to be nursery reared.

Being a believer in visual aids when trying to convince someone of your point of view, I listed the unconfirmed nursings using highlighter pens of different colors, noting the individual behaviors that we suspected were culminating in nursings:

Green—infant antsy and sometimes vocal
Pink—Molly shifting infant and turning her back to keepers
Yellow—infant quiet for a consistent period of time
Orange—Molly turning back to keeper with a sleeping infant
     in her arms

The time between sequences of events was consistent with the length of suspected nursings—all excellent indicators that nursings were occurring. What emerged on paper was a kaleidoscope of green, pink, yellow, orange lines, each indicating a distinct and repetitive pattern and timeline. This gave all of us, including veterinary staff and upper level management, enough confidence to leave the infant with Molly. In time, Molly became more comfortable with allowing the staff to observe the infant while nursing. It was the strength of these behavioral patterns that bought us the much-needed time to allow Molly and the infant's bond to strengthen and grow.

Molly & Infant Nursings 15 November 1992 (1st time mother)

Day 1: 15 Nov 1992

8:43 a.m.	1:00 minute	Left
9:28 a.m.	1:40	Right
12:35 p.m.	2:25	Left
1:25 p.m.	1:50	Right
2:30 p.m.	1:00	Left

Total: 7:55 minutes (5 nursings)

Day 5: 19 Nov 1992

8:15 a.m.	1:40	Right
10:10 a.m.	1:50	Right
10:13 a.m.	0:30	Right
11:50 a.m.	2:57	Right
1:27 p.m.	1:30	Left
3:40 p.m.	0:40	Right
4:56 p.m.	0:45	Left
4:58 p.m.	4:00	Left
6:14 p.m.	0:20	Right
7:01 p.m.	0:15	Right

Total: 14:07 (10 nursings)

Day 8: 23 Nov 1992

7:00 a.m.	1:30	Right
8:40 a.m.	3:00	Left
9:50 a.m.	1:10	Left
9:53 a.m.	4:15	Left
9:58 a.m.	1:30	Left
10:45 a.m.	2:45	Left
10:49 a.m.	0:25	Right
11:00 a.m.	3:00	Left
12:10 p.m.	2:50	Left
12:30 p.m.	1:35	Left
2:10 p.m.	2:00	Left
2:15 p.m.	2:15	Left

3:50 p.m.	1:45	Right
4:12 p.m.	2:10	Left
5:00 p.m.	2:05	Right
5:10 p.m.	1:30	Left

Total: 35:10 (16 nursings)

44

SATURDAY NIGHTS

I work the late shift every Saturday night during the summer months, closing up the Ape House at eight. I also work late on our zoo members' nights on Wednesday, but because Saturday is my last workday before my weekend begins, I feel a sense of relief and look forward to going home.

Like clockwork, around 5:30, the frazzled-looking mothers and fathers take their exhausted, sunburned children home after spending a day at the zoo and water park, but a few stragglers remain. The gorillas in the habitat know it's going to be a long wait until I can start moving them in at 7:15. Some are antsy, some resigned to the delay in getting on with their nightly routine. It must feel a little bit like when you are a kid at an all-day family function and you're tired, ready to go home to your own space, to your own bed, to your own routine where you can truly relax but must instead wait on the schedule of your parents.

I clean the kitchen, stock shelves, and fill the food bins with cereal, popcorn, seeds, and biscuits. Earlier in the day, my fellow keeper and fellow clean freak, Adele, and I did our usual Saturday cleaning routine. Every Saturday, we do a thorough bleach and scrub of everything we can get our hands on: gorilla enclosures, climbing structures, food bins and bowls, the gorillas' drinking cups, kitchen cupboards, and food scoops. In the afternoon we borrow a zoo vehicle from another department, weave our way through the weekend crowds up to the zoo barn to load up on bales of hay, stopping by the diet kitchen to add bags of mixed seeds, monkey biscuits, and industrial-sized boxes of Cheerios before heading back down to the Ape House. When we get back, one of us will pop popcorn to be added to one of the large gray trash barrels we have at the end of the back keeper aisle used to store the dry treats. By the time Adele leaves for the day, the Ape House smells fresh and clean, with a slight whiff of bleach and sweet alfalfa hay underneath the gorilla smell.

It's still too early, so I don't want to prep their indoor spaces with fresh hay yet because the gorillas will hear the doors opening and closing. It will false start their expectations that they will be coming in soon. It would be an unkind tease, so I wait.

I grab some frozen treats from the freezer. The gorillas had treats earlier in the day, but screw it; it has been an incredibly long and hot day. I dip the bottoms of each eight-ounce plastic container of frozen fruit yogurt into a bowl of hot water to loosen them, and the solid block of yogurt easily slides out into the waiting treat bowl. They are popsicles minus sticks.

The gorillas see the big blue bowl as soon as I walk out the keeper door to the habitat. The boys, Mac, Mosuba, and JJ, hightail it over so they can get theirs before anyone else. They grab their treat and quickly lope away, looking over their shoulders as the adult male Mumbah approaches. The twins and JJ look as if they have contraband, as they scramble up the mesh with their treasures, up and away from potential trouble.

Mumbah shambles over, takes his treat, and elicits a half-hearted cough-grunt when he sees me getting ready to give an adult female one of the yogurts, so I move off—the female, moving quietly above me on the mesh, follows.

During distribution of food, the gorillas, especially the low-ranking ones, count on the keepers to assess their locations and quietly but efficiently get them their food, with little fanfare and without drawing any undue attention to them.

These treats are so valuable that I have seen juveniles scream in sheer frustration when an adult suddenly grabs the treat out of their hands. If that happens, I head back to the kitchen to get them another, and when I next come out, the treat is safely tucked hidden in my hand. The youngster sees me as I move through the pampas grass surrounding the exhibit acting as if I am on some other unrelated mission. He or she follows my every move, finally meeting me on the far side of the exhibit where the covert handoff occurs. They depend on us to deliver, to watch out for them. This solicitousness is not to be confused with interfering in their lives but rather reassures them that we see them, that we see their needs and will respond accordingly. We have their back when necessary but then get out of their way.

It is quiet. The evening has started to cool slightly as the shadows lengthen. Outside the perimeter fence is Dury Sudduth. Dury is a long-time docent, and every Saturday either he or another docent, Mike Moore, works the late shift with me. After extensive training, a Columbus Zoo docent's job is to interact with the public. They answer questions about the gorillas sometimes sharing moments they have observed while on duty. What is highly unusual is that our docent organization is quite independent and as is typical of our loose management style not under the thumb of an administration. Because of that, the docents are exceedingly creative and productive in their own right.

There are no people to interact with at the moment, and I'm still a good hour or so away from bringing the gorillas in, so Dury and I stand on either side of the public perimeter fence—my right foot comfortably propped through the mesh, my arms dangling loosely over the top rung, discussing the day. There are certain docents who a keeper can rely on, they don't presume, they don't embellish or interpret; they just report what they have seen. Most importantly, they are discreet; they don't share my concerns with the public. Both Dury and Mike are like that. If something is going on in the troop in the outdoor enclosure, I can go to either of them, voice my thoughts and concerns, and ask them to keep an eye out with the confidence that they will report back concisely without exaggeration, thus allowing me to get on with my other keeper duties. I depend on them.

Some of the gorillas wander over where Dury and I chat. Cora stops by. She sits sideways to us, stretching her legs out straight along the habitat's narrow cement foundation wall. She appears to listen in on our conversation but keeps a close watch with frequent glances over her shoulder for any approaching troop members. One of the boys meanders over to see what's up, or to try and engage Cora in a bit of play.

Colo comes over next, displacing the lower-ranking Cora. Sitting with her belly flush against the mesh facing us, waffling her belly pelage. She sends loud raspberries our way to get our attention. If that doesn't work she resorts to very neat sprays of spit—think of the gap-toothed kid in your neighborhood while growing up who could spit the longest and most precise stream of water ever. If that doesn't get our attention, she may gather a few discarded branches from the enclosure, break them into small pieces and toss our way, with great accuracy.

One of the boys wanders over our way again and then gallops off to join the other two who are playing king of the hill, chasing one another up and then down the hill and through the tunnel underneath. Their attention span at this age is short-lived, and of all the gorillas that have to endure this late shift every week, it's the juveniles who don't really seem to mind too much. They strike me as being somewhat similar to the solid metal balls in a pinball machine. They move from one potentially interesting situation to another—bing, bing, bing. Or perhaps even more like a group of kids on a Saturday night playing hard into the late evening, kids who don't have to come in until dark and only reluctantly head home upon hearing their father's whistle or their mother's distinct voice calling for them.

Dury and I discuss anything and everything, including changes in the gorilla groups. I share behind-the-scene stories of the gorillas. Then we talk of world events, but with Dury being a true Anglophile, we often discuss the UK and his frequent travels there. And because I have been reading books about English medieval court intrigue since a young girl, we discuss English history.

The shadows are getting longer, so I know it's time to start prepping. I have already done their individual food bowls and cut up their greens so I head in to add extra hay to each room. I fluff it up and then spread their greens, seeds, and cereal throughout. The gorillas are antsy now, anxious to eat and settle in for the night. When they hear the clang of the heavy doors shutting, they begin their feeding vocalizations, each wholly distinctive.

I open the air-compressed doors in the overhead chute system giving them access down. As they come in, they will divide themselves out as to where they want their individual meals served. Again my job is to make sure no one gets caught in a tight space with another gorilla. Emotions run high at feeding time and fights can break out. Because the long chute system has numerous doors that can be broken down into smaller spaces, the gorillas often choose to sit up in the chutes to eat their meals. Every gorilla has a big blue plastic bowl with their whole fruits and veggies—sweet potato, carrot, celery, apple, lemon—which are handed individually to them by the keepers. And everyone has a sixteen-ounce plastic cup with their name on it for their three times a week protein drink and their daily juice drink. We don't give the cup to them directly but rather hold the cup up to the mesh and they drink from it as we angle it ever higher so they get every last drop.

Once they have eaten, there is a method to letting them back together in their communal sleeping quarters. The whole group will be placed together overnight, but it's imperative to let the juveniles and females down from the chute first, get them settled in for a few moments of uninterrupted browsing for their cut-up greens, and then add the silverback last. If the silverback were to be let down first, and I was still in the process of moving a female down from the overhead chutes, a male could climb on the stoop essentially blocking the chute door and trapping the female, reaching his arm in, causing her to panic. Once that happens, it can have an escalating effect. She would start to cough-grunt, which in turn pisses the male off. When those emotions are heightened, I have seen much larger males surprisingly fit through a smaller-sized "female" door when incensed and then all hell can break loose. Serious injuries can and do occur when we keepers make mistakes.

I was impatient one day while transferring animals and didn't take the time to assess the situation properly, and because of that, during an altercation with another gorilla Mac (the twin) was injured, resulting in a deep wound to the bottom of his foot. I felt just awful. It was my screwup but he was paying for it. Every day for several weeks he limped his pitiful self over to us, propped his foot up on the metal framework so we got a good look-see while he scanned our faces, looking for sympathy it seemed. Then we commiserated with him, fussing and tut-tutting over his wound, squirted hydrogen peroxide on it and gave him a treat before he gamboled off to his troop.

Introductions were always touchy and if not planned carefully injuries could occur. Molly had moved over from Oscar's building, and we had begun the introduction process to Bongo and Fossey. But because Molly was somewhat immature, she was given to displaying frequently around Bongo, much to his annoyance at times. Returning from my day off, I learned from Dianna that Bongo had injured Molly. But before I could open up the solid keeper door to the back aisle she said, "I have to warn you, it's bad." "OK," I replied, kind of shrugging it off, but truthfully nothing could have prepared me for it. When I opened the keeper door, Molly was sitting at her back enclosure door and I had a clear view of her wound. "Holy crap," I exclaimed. Her inner arm was laid wide open, with a deep wound running about six inches long, starting just above her wrist and ending just below her elbow. She calmly looked at me while gingerly cradling her arm.

Often there is a reluctance to close up a primate wound with stitches for fear of an infection setting in—so it was decided to leave it open. We spoke to the vets and convinced them to let us try an alternative treatment. We loaded her up with extra vitamin E and C every day, gave her additional protein drinks, and flushed out the wound two to three times a day with a large syringe of warm hydrogen peroxide. We proceeded with our plan and amazingly, within six weeks the wound had completely healed over.

After the gorillas have eaten, been given their drinks, and have been put back together, I ask Dury, "Do you want to come in and sit for a while to observe?" Opening the gate for him, we enter the building, sitting on the front bench. The gorillas don't mind, they know Dury from his frequent docent stints around the gorilla habitat, and besides, they are all too busy mucking around the hay looking for the unseen goodies hidden within. Some of the females have already started gathering their hay to build their night nests. We sit quietly in companionable silence just watching, taking it all in.

45

BONGO AND FOSSEY

It is late August 1990, just a week past Fossey's fourth birthday. At thirty-four years of age, Bongo has been in captivity for thirty-two years. The morning cleaning is almost done, the Ape House smells of wet cement and fresh hay, and the soothing sounds of Mozart play on the tape deck. Bongo has chosen to stay indoors. He has access to the small outdoor enclosure, but being Bongo, he opts out of being on public display. He likes his quiet, his solitude, but he is not alone; his four-year-old son, Fossey, is scampering in and out of the building. We have all decided that Fossey is the best. His presence was a watershed moment in the gorilla husbandry program, and he is so much more than just a young gorilla. His birth and subsequent rearing by his parents solidified our gorilla husbandry program and reputation

After Bridgette's death, Bongo immediately took over the role of sole caregiver to his son without a hitch. Other females were eventually introduced, but the bond that these two males have is as essential as breathing. They are buddies, partners in crime, pals, father and son. Their story of kinship is one of such infinite beauty that the keepers love coming to work every day just to see what this pair will do.

Today on this late summer day, Bongo sits quietly on the floor, his legs outstretched in a perfect V, back ramrod straight. Bongo vocalizes to Fossey as he comes running through the short chute that connects the small outdoor area to the indoor one. Fossey swings down, using the metal handholds on the wall to make his grand entrance

FIGURE 45.1. Fossey and Bongo playing and laughing

into his father's space. He is energy in motion, very busy today. He runs over to his dad and rolls into him, lying flat out on his back, arms and legs spread, belly and groin exposed, in invitation for a tickle. His dad does a perfunctory "naa-humm" vocalization, looks down at his son in a stately sort of way; he waits a moment before reaching out a huge hand to tickle his boy.

Fossey happily laughs, then his laugh turns into a gasping chortle as his dad leans forward, holds his son in place, and continues to tickle him while nuzzling Fossey's belly with his mouth. There is such trust between them, this four-hundred-pound adult male and his seventy-five-pound kid. It would never occur to Fossey that his father could or would hurt him. They absolutely adore one another. As quickly as it began, it ends abruptly, with Fossey off again, running, swinging through the enclosure to the outside.

The next time Fossey comes inside, he sees his father walking along the horizontal floor-length climbing structure. The wood structure is similar to a balance beam, raised several feet off the floor, but this beam is slightly wider. Fossey decides this is an excellent game and sits underneath his father, poking his feet as Bongo passes over him. Soon they are both rumbling to one another and then Bongo elevates the game by repeatedly running back and forth. Fossey runs bipedally beneath his father, laughing while grabbing at his dad's feet above. Bongo looks a bit like the cartoon hippo sans tutu, balancing on a narrow substrate doing silly twirls on delicate feet.

I am in the adjacent enclosure, lost in the soothing monotony of hosing the floor. Hearing their mingled deep-throated laughter, I stop, kink my hose, lean against the mesh and watch them. They are oblivious to me and I revel in my privilege, of bearing witness to such deep devotion.

Several weeks later, I was early for once, the first to arrive. It was such an anomaly for me, a person who struggles with early morning, with being on time. Surely it was made to happen, like so much else in life—a change in plans, a missed appointment, a detour that puts you right where you need to be at the very moment when you are essential either as a witness or a participant. It was a gift of perfect timing, of circumstance, dropped in front of me. And once there, why did I open the side door off the kitchen first, which was usually the last place to be checked? But I know why: Bongo was there.

We all come to our calling, our passion, in different ways. Some are sure and intent on how they will get there, knowing from a young age they want to be a doctor, a nurse, a teacher, or a veterinarian. But others of us who are less certain, less confident in our abilities, come to find our place in this world through what seems like unrelated and circuitous routes. Until one day we finally arrive, breathing a sigh of relief saying to ourselves, "Ahhh yes, this is where I belong." From an early age, I found animals fascinating. After church on Sundays in late spring, I spent hours flat on my belly in our driveway, arms crossed, chin resting on top, observing columns of ants coming and going. I fed them tiny crumbs and then watched in fascination as they struggled to pick up and carry their gargantuan prizes. They labored along drunkenly until they found their balance, and then they became focused and seemingly unstoppable. Sometimes ants coming from the opposite direction stopped, ostensibly to speak to the crumb carrier, like a couple of elderly people in a small village out on their daily walk who briefly stop to exchange pleasantries and gossip and move on toward their own lives and daily obligations. Some of the ants that stopped seemed a bit abrasive, pushy even with my crumb carrier. They gave the carriers a once over in an aggressive manner. Were they soldiers perhaps, the keepers of the gate? Thus began my fascination and predilection for observing animal behavior.

Bongo is slumped forward, his face resting gently on the concrete floor. It was obvious he had been in his usual position, sitting straight up, his legs at a V in front of him. It is as if he were a tree that had just been felled. His four-year-old son, Fossey, moves tentatively around his father's body, his uncertainty evident; he is perplexed. Every few minutes Fos leans down, looking into his father's slack face, and then quietly begins circling Bongo's body again.

Later, I would wonder about that day, the timing of it all. And all I will feel is a wave of thankfulness that I had been the one to find Bongo, to bear witness to his death without the company of others. To see his young son bending down, his tiny bum in the air, his worried face inches away, peering into his father's unseeing eyes, while quietly and desperately hooting for some measure of a response, before I say, "Ohhh, Bongo," and begin to weep.

Fossey is quickly integrated into another gorilla troop, where he is a friend with JJ who is his own age. The group is made up of multiple females, some juveniles and youngsters under the age of five, and Mumbah. Adult female Sylvia was also transferred to this new troop with Fossey, as she had briefly been a part of Bongo's group. She is somewhat an adoptive mother to Fossey, but she is not so invested and a bit careless in her relationship with him. He is not an infant that needs full-on nurturing, but neither is he a fully-fledged juvenile yet. He is in that no-man's-land, needing comfort and some protection from an adult but also developing his own independence. It was his father's presence that gave him his sense of self, his sense of confidence.

The staff watches Fos closely to see that he is adapting. During the day he seems fine, he plays hard, he and JJ creating havoc in the troop. JJ has been raised within the group, having been adopted by Colo so he belongs, he knows his place and is confident, even brash in this knowledge. Fossey is more retrospective, more cautious. He knows he is a latecomer to this troop, but he appears to be settling in.

Several months after Bongo's death, I am on the 10 p.m. to 3 a.m. shift, observing Lulu, who has given birth to her daughter Kebi while in Mumbah's troop. The Ape House is in semi-darkness when I arrive. Grabbing a flashlight, I walk down the back aisle to the front of the enclosure. Lulu has chosen to nest up in the overhead chute with her newborn daughter. Almost hidden in a fluffy wood wool nest, she lies comfortably on her side, her baby sleeping. She rumbles to me in greeting when I whisper, "Hey, Boo, how's your girl?" Then she settles down to a nap before her next bout of nursing. Mumbah's troop is asleep, gorillas are scattered throughout, some on the floor, some in transfer chutes, all bedded down in hay nests. Sylvia is settled on a small cement bed raised several feet off the floor—her great bulk taking up the entire shelf. Fossey, at four and a half on this February night, is wandering throughout the exhibit. He has no place to sleep, as Sylvia's narrow bed does not afford him any room. He's looking to settle next to a warm body, to be lulled into sleep by the breathing of another gorilla, to feel the safety and comfort of arms around his young body. He wanders with his right arm across his chest—an indication since infancy of insecurity, as he softly hoots to himself, searching to settle somewhere.

Lulu nurses her infant on a regular basis; she's a good mom. I note the time and length of each nursing, the infant's vocalizations, and the respirations when at rest or awake. Everything looks good. But it is Fossey I keep coming back to. He settles down for a bit, before getting up and wandering, his hooting a haunting accompaniment. He is a forlorn figure, quietly and desperately in search of his place, looking for where he belongs.

46

MUMBAH AND PONGI

Our surrogate dad, Mumbah, has always had a faintly vacant look to his face, a lost look in his eyes, his lower lip hanging slack. He is smallish, compact in build. He doesn't have the "wow" factor that can emanate from many adult male gorillas.

Having fathered an infant while at Howletts, he was brought in as a potential breeder, but over time it became clear he showed no real or consistent inclination to breed with any of our females.

Because of Mumbah's exceptional gentleness and to some degree his lack of assertiveness, we thought that we could safely integrate unrelated infants into his troop, with Mumbah "adopting" them as his own. Mumbah did not disappoint. The adoptions began in the spring of 1988 with the introduction of fourteen-month-old JJ and continued on with numerous infants integrating into his troop. But in 1993, another event involving Mumbah just took our breath away.

Tragically in late June, Pongi's twenty-four-year-old mate, Oscar, died after a physical examination, leaving behind his six-year-old son and a very pregnant Pongi. As happens too frequently, animals can have a difficult time waking from anesthesia. In this case, Oscar had suffered a heart attack. Both she and Oscar had raised their son, Colbi, together and were excellent parents. At the time of Oscar's death, Pongi was quite near her delivery date.

Pongi had come to us in 1985. I was working in the Ape House kitchen one morning when the phone rang. It was the Birmingham

Zoo and they said they had a female they were looking to place. Then the caller, the curator or director (I can't remember which), moved onto the regular litany of complaints. His description was indicative of the times, especially coming from upper level management; she was too old, wasn't interested in breeding, was not socially sound, was housed with their male Joe but they did not get along. I asked how old she was and to my shock, he told me nineteen, and he again reiterated his view that she was too old to reproduce. I immediately said I would pass along the information and have someone get back to him ASAP. Have you ever had that feeling when you know you have just hit on the ultimate prize and no one else knows at the moment what a treasure it is? Nineteen was young. Pongi was not too old to breed, let alone reproduce, most likely she just wasn't housed in an environment that was positive for her and her mate. I went to Dianna, "You're not going to believe this." Dianna then went to Don Winstel, and they very quickly got back to Birmingham telling them we would be happy to take her. She was a blessing and would prove to be an extraordinary gorilla. Years later when Randy Reid, Pongi's keeper at Birmingham, came to Columbus for a visit, we talked extensively with him about her move to Columbus. Randy, of course, had never felt that way, had never laid blame at her feet. He knew her and he knew gorillas.

I came back to the gorilla department as the head keeper just a few weeks after Oscar's death, after spending the previous fifteen months as the head keeper of the Australia/Asia area. Having few options, we decided to try and get Pongi integrated into Mumbah's troop prior to the birth. This meant that we were fast-forwarding our introduction protocols in the hope of having the intro accomplished by her due date. A successful introduction was defined when the new member was spending the night with all group members 24-7, including the silverback male. A couple of years prior, we had introduced Lulu while pregnant, a highly unusual and provocative step—some would say a dangerous one. But from that experience, we knew Mumbah would most likely accept another female giving birth in his group without harming the infant.

On the morning of August 15, 1993, while doing my early morning rounds, I found Pongi had given birth ahead of schedule to what appeared to be a healthy, thriving infant. Our dilemma was that

Pongi was fully integrated in with all the females and their various offspring (biological and adopted), but she was not as of yet with Mumbah twenty-four hours a day. The entire troop, including Pongi and Mumbah, spent five hours a day together outside. Our next step in the introduction was to leave them together overnight, but obviously we had either gotten her due date wrong or she had delivered early and we simply ran out of time.

So although Mumbah would have witnessed the birth as he was housed next to Pongi, we could not be absolutely certain how he would react until we put them all together that day. The keeper staff felt pretty certain he would tolerate this new addition to his troop, but that certainty was seasoned with loads of caution. We quickly cleaned the outdoor habitat, prepped it with a large variety of food and enrichment items in order to keep everyone occupied and distracted, and then we let them all outside.

Pongi was a smart female, sophisticated in her social approach. She was neither submissive nor overtly fawning toward males. She was a female that knew how to work a male, quietly, subtly, and determinedly. From the moment she went up into the long chute system to climbing down into the exhibit, Pongi took charge. She was on a mission—it was in her demeanor, bearing, and body language, and we heard it in her continual low vocalizations. With her newborn infant cradled in her arms, she made a beeline straight toward Mumbah, who was sitting relaxed in his usual spot in the habitat.

When she got to Mumbah, she stopped directly in front of him and stood still as if presenting the infant. She did not hold it out per se, but somehow she gave us the impression she was showing him the infant. Pongi was looking for some sign of recognition and acceptance from Mumbah. Whatever cue he gave her, it was so subtle we could not discern it, but something clearly passed between them, because she seemed satisfied, relaxed her stance, and promptly sat down very near to Mumbah. From that moment on, the entire troop was together twenty-four hours a day, end of introduction. Pongi went on to be the dominant female and raised her daughter Casode (Cassie) within the troop. In turn, Cassie, sixteen years later, would give birth to her own daughter in the very same enclosure of her birth, and successfully raise her kid, Pongi's granddaughter.

47

SOCIAL BEINGS

Like all primates, ourselves included, gorillas have social parameters, a structure in which they navigate their daily lives. They have rules that govern acceptable behaviors. These social constraints act as a check and balance on how they conduct themselves in relation to others in their group. Signals are sent, some obvious, some less so, but for the social rules to work they must signal that they can be trusted; the signals they send must be true, not ambiguous or deceptive.

In 1993 things started to change at the zoo. The gorilla keepers had always played an essential role in deciding where the program was heading. They were in on any and all discussions concerning sending our gorillas out or receiving new gorillas. So it was with some shock that, while at a national zoo conference, I learned from the Philadelphia Zoo curator that Columbus would be receiving two brothers from them, ages eight and nine. I was both taken aback and alarmed on several levels.

At the time, I was in the Australia/Asia section as the head keeper, so in reality it was none of my business. But as I was still in close contact with the gorilla keepers, especially Charlene, I was surprised I had not heard anything. Equally concerning were the ages and the plural nature of it. Eight- and nine-year-old males are not adults. Some may look like silverbacks, but they are not. They are blackbacks, and because of that they are in essence still teenagers, albeit large and physically imposing teenagers.

Two events played into all of this, I'm sure: twenty-four-year-old Oscar's untimely death earlier in the summer and the recommendation and push on the part of the Gorilla Species Survival Plan (SSP) to get new bloodlines going in the North American gorilla population. Our males Sunshine and Oscar had both been prolific breeders, as had several other males throughout the North American population, so certain genetic lines were currently overrepresented or projected to be in the future.

Although the SSP can make recommendations, a zoo can decline to receive animals for a number of reasons: space issues, overcrowding, and staffing shortages. In the past, we would have put the brakes on a firm commitment in order to go back to discuss these recommendations with the Ape House staff. But according to the Philly curator, it sounded like a done deal. When I got back to the Columbus Zoo, I walked over to the Ape House and asked Charlene, Dianna, and Adele if they had heard of any of this. They had not.

In order to make room for the Philly boys, Anakka and Chaka, the zoo decided that our own Fossey and JJ would be shipped to the Little Rock Zoo. If they had to go anywhere, this zoo with its like-minded keeper staff, headed by Ann Rademacher, was the perfect place for them. Little Rock shared our philosophy. Most importantly, their management truly listened to and empowered their keepers. Fossey was seven and JJ just shy of seven. Because they had formed a close bond while in Mumbah's troop, it was thought that they could co-lead a troop in the future, and moving them at this age would allow them the time to develop their roles before becoming fully mature males.

Also, Fossey had a gentle approach to others, which boded well for his leadership skills. Who knew the source? Perhaps it was because he was raised by his father and losing his mother at such a young age had forced him to adapt. Fossey had few of the tendencies of males his age, such as pushing to see how far they could go. Maybe being displaced after his father's death somehow shaped a more temperate soul; he had an easygoing, tolerant manner. At the risk of sounding completely anthropomorphic, also recognizing that he had my heart as his father had, he was a gorilla that I would say had empathy.

*   *   *

The Philly boys were unpredictable. We initially tried them together as a team with some of our females to see if they could co-lead. But unfortunately, and most likely a product of their young age, they proved to be formidable and erratic.

I remember seeing Anakka sitting quietly across from a nervous Toni, who had been pulled from Sunshine's group to try to form a new troop with Anakka and Chaka. Again, this was so against the grain of our program. Toni had not given any indication that she wanted to leave Sunny's troop. We were trying to force a square peg into a round hole, always a recipe for disaster with captive gorillas.

Anakka and Toni were sitting face-to-face, extremely close, as if having a casual chat. Anakka's face was soft, his body completely relaxed—no indication of tension or aggression. He made a nice vocalization to Toni and although still nervous, she seemed to let her guard down for a moment—and then he sucker-punched her in the belly. No warning. She didn't see it coming, and I didn't either. Shocked and startled, she quickly recovered and lashed out, screaming and then chasing him. I got the distinct impression that's exactly what he wanted—a reaction. She was frightened and incensed, actually shaking with outrage. You can imagine the tension that this altercation caused within the building with Sunshine tucked away in a back enclosure. He was going crazy hearing her screaming; he was frustrated and absolutely furious. That is pretty much how the introductions proceeded with these two blackbacks, causing untoward stress to all in the building.

Another incident occurred when I was gone. I was in Florida at the time, visiting my grandmother, when I received a call that Cassie, Pongi's young infant, had a serious injury. Her thumb had to be amputated because apparently sometime during the night someone had bitten it, practically shearing it off. We never knew for certain what had happened but can only surmise that Cassie may have been climbing up the mesh when either Chaka or Anakka bit her hand, as they had been housed in the adjacent enclosure.

Eventually they were split up. Chaka went to the Cincinnati Zoo, and we integrated Anakka into Mumbah's troop. We did so reluctantly but hoped that a fully mature male's presence might serve as a source of discipline and an example of good leadership skills.

There was a pretty famous story that went around in the 1990s describing the role of older males in socially complex societies guiding youngsters. South Africa routinely culled their elephant herds back then—the thought being the populations were getting too large and as a result they were decimating ecosystems in the National Parks. But by removing adults, the juveniles were left on their own. Think *Lord of the Flies.*

For an animal as socially complicated and intertwined with one another and as intelligent as elephants, the trauma of what the survivors witnessed would have been analogous to us seeing our own family members murdered in front of us. Carry that shock, anger, and emotional pain with you as you move forward and combine it with no adult supervision, no steadying influence or comfort, and you have created a volatile foundation for out-of-control behaviors in these highly social animals.

The results were that teenage male elephants formed gangs, destructive gangs that went after other wildlife (endangered rhinos for one) and tourist vehicles. They had turned into cock-sure, unchallenged, and dangerous bullies. So guess what the solution was? They brought a mature bull elephant into the juvenile's territory and into their daily lives. The adult male's presence alone was what these young adolescent males needed, they needed guidance and social parameters enforced. If they strayed or pushed beyond those boundaries, they were reprimanded and brought to task by the bull. Not surprisingly, it worked.

Mumbah's tolerant, if somewhat removed attributes that made the surrogate group run smoothly and without drama did not necessarily play out effectively when reining in an unruly teenager. Mumbah was not an actively forceful silverback, and handling Anakka was way above his pay scale.

We do a huge disservice to troop members of all ages as well as the blackbacks themselves when we ask immature males to become leaders prematurely. Not only are these males less wise in their interactions with fellow gorillas, but for the most part they haven't yet developed the subtle nuances that can entice a group member to their side. They aren't often even likable. But in their defense, they

are simply acting as most teenage boys do: exhibiting annoying behavior, poking a finger at those around them, just hoping for a reaction, the more extreme the better. Anakka wasn't doing anything he
should not have been doing at that age; he was testing his boundaries,
pushing and pushing some more to see what he could get away with.
Our mistake was asking these juveniles, these blackbacks, to have the
knowledge, the wisdom, and the restraint to take on the mantle of
leadership.

Eventually Anakka was given three females, Colo, Toni, and
Jumoke, to form his own troop, but he continued to exhibit these
same disruptive behaviors. Initially, the females did exactly as we had
hoped; they banded together and backed one another up when one
was being picked on. But as time went on, I can only say that our observations indicated that the females were worn out by his need to get
a reaction all the time. On any given day, a particular female would
be targeted and the other females that were not being targeted at the
moment seemed relieved that it wasn't them that day. So they subse-
quently began to keep their distance, staying on the periphery, each
reluctant to come to another's aid. This sort of brutish behavior was
being reinforced on a daily basis. In essence, it was teaching Anakka
that this was a perfectly acceptable way to lead a troop. We weren't
doing anyone any favors, including Anakka.

When Anakka's first kid was born in 1997, Charlene and I were
asked to come back to the Ape House to observe his behavior with
his newborn son. By that time we both were working full time on
field conservation for the zoo. I was the field conservation coordinator, overseeing and distributing funds to field projects around the
globe, and Charlene was the cofounder of Partners in Conservation,
working with local communities and people of Rwanda in support of
mountain gorilla conservation.

We sat on our familiar front observation bench with keepers
Susan White and Debbie Elder. It was like old times. Our philosophy had always dictated that males should remain with their females
during and after a birth, but that was completely dependent on the
male's appropriate behaviors. In this case, Anakka was separated but
was able to see the birth so he would have known it was his offspring.
We watched as the whole group was put back together, and sure
enough, Anakka's shenanigans began. He started pursuing new mom,

Jumoke, wanting to touch the baby. He was insistent, and she was nervous. Jumoke's appeasing vocalizations started low on the scale but soon began to escalate into a higher pitch as she started to panic. Anakka was relentless, following her everywhere, as she placed a protective hand over her infant who was clinging to her lower belly. I don't think he wanted to harm the baby, but he wanted to touch it, and Jumoke clearly was not comfortable with that. The other females hovered nearby, also wanting to protect the baby. There is one thing that is abundantly clear when an infant is born: it is the mother who gets to call all the shots, deciding who gets to touch the baby, when they can touch the baby, who needs to keep their distance, and who is allowed proximity. Anakka was clueless, and the stress was telling on Jumoke's demeanor.

Our general conclusion was this. It was Anakka's genetic material, his child, strapped to Jumoke's belly. Anakka was supposed to be a comforting and protective presence in this first-time mother's life, so that she could relax and raise the baby in peace. He was none of these, so we suggested removing him until he could prove that he could be a distant but doting dad, respecting the boundaries that the mother established. Anakka's good behavior was nothing more than an investment in his future, but he seemed unable to grasp that.

Observation Notes: Anakka's Third Kid, 2010

Infant (born 10 September 2010)
Mother: Cassie
Father: Anakka

2 a.m. & 4 a.m.—Anakka pursued Cassie (4:00 min. Really startled her).

8:13 a.m.—Anakka separated from females after 5 minutes of pursuing Cassie. This was a change in intensity and intent. He wanted to see and touch baby and was insistent about it. Cassie's personal space was clearly not respected. She was very nervous/frightened.

12:12 p.m.—Group all back together (Cassie had a nice rest for a few hours). Anakka and Cassie very close together, Cassie appears pretty relaxed, on her back, resting, Anakka foraging nearby.

12:37 p.m.—Anakka done with foraging, starts pursuing Cassie.

12:55 p.m.—Anakka displaces Cassie from her mesh bed.

12:59 p.m.—Anakka displaces Cassie from her chute bed several times. He's relentless, Cassie very uncomfortable.

1:15 p.m.—Anakka displaces Cassie from her mesh bed. He pursues Cassie.

1:20 p.m.—Keepers separate the gorillas.

48

WHY WE DO CONSERVATION

An enduring memory—

Bongo is sitting in one of his typical Bongo poses, hunched over on his belly, his weight resting on his elbows, leaning forward, his arms crossed in front of him one hand holding his opposite wrist. On the floor in front of him is his collected pile of sunflower seeds. He bends over slightly, and delicately picks up one seed at a time with his dexterous mouth, de-husks each with his lips, swallowing the seed within. Then Bongo efficiently spits out the shell.

Another image pops into my head, from my early years in the Ape House. It is winter, and the building is extremely warm. As with humans during cold winter months, when the furnace is running nonstop, the brittle air has a tendency to dry out the gorillas' skin, making their hands and feet crack. Dianna has given Bongo a cotton cloth soaked in baby oil. I watch as he does what we humans do. He rubs the cloth over his hands, front to back, using it like a washcloth. He goes from one hand to another before finally walking over to his water bowl where he soaks the cloth, wrings it out, and then proceeds to goof around with it, tossing it in the air and expertly catching it.

Bongo's individual idiosyncrasies were his and his alone; they represented how each animal has a distinctive personality, which translates into unique stories. Shortly after I began working with Bill and Dianna, I was asked to come back to the zoo late in the evening to check on a cheetah Bill thought was close to giving birth. After peeking in on her—no sign of labor, no babies yet—I decided to walk the short distance to the Ape House. I'm not really sure why, maybe I was just curious. What were they up to at this time of the evening? I opened the double doors and walked into the wide expansive public aisle (bear in mind this was pre-renovation so the building was still open to the public), taking a seat on the front bench and facing Bongo's enclosure. The lights were off in the public area, but the Coke machine cast a soft reddish glow along the shiny concrete floor. The keeper aisle lights behind the viewing windows were off, but above each of the five enclosures the lights were still on, dimmed but on. The reasoning was that it enabled the security staff to see the animals when they did their nightly rounds. Bongo was lying on his back on the bare cement floor, his head propped on the four-inch raised cement water bowl, using it as a pillow of sorts. He shifted frequently, searching for a more comfortable position. This massive beautiful animal looked completely exposed and vulnerable, and I felt shame wash over me.

There was much discussion in the zoo world in the 1990s around the most effective approach to supporting conservation: the ecosystem versus individual animal argument. Bear in mind this was just talk. It was, for the most part, theory, nothing more. Many, if not the majority, of zoos were still not supporting in situ conservation in any meaningful way. In a nutshell, many thought the ecosystem standpoint was what we needed to concentrate on, the big picture stuff, the overall habitat. A focus on the individual animal or groups of animals was thought to be too narrow of an approach. I remember getting into a disagreement with one of the better-known names in the zoo world that dismissed primate sanctuaries in Africa as being ineffectual, draining resources that should have been going into habitat protection. My counter argument was that these sanctuaries, that cared for confiscated young gorillas and chimpanzees, served not only as safe havens for these injured and traumatized animals but also as important education centers. These sanctuaries also demonstrated a cooperative relationship with local and regional law enforcement within that particular country. Equally important is that each of these animals had a story to tell that could move people to positive action.

And I might have taken his viewpoint more to heart if zoos had been actually supporting much of anything in the field. But other than the Wildlife Conservation Society, which was the undisputed leader, Busch Gardens, Columbus Zoo, and less than a handful of others, I did not see a substantial conservation commitment from the zoo community then.

This was the same sort of thinking that I believe informed the decades-long approach to studying animals in the wild, i.e., number your study subjects be they chimps, elephants, or gorillas. Don't give them names. Giving names may make the relationship too personal, not objective enough to be true research. The field researcher might go down that slippery slope of seeing personality and anthropomorphizing their study subjects, seeing them as individuals with very real personality traits. Yet in the 1960s and '70s, field researchers like Jane Goodall and Dian Fossey began naming their study subjects, famously David Greybeard and Digit, respectively. And with that gesture came the inevitable personality traits that enthralled the public. We learned that conservation is personal, and what these animals endured was personal and because of that their individual stories were extremely powerful tools to raise support and awareness.

In 1990 the Columbus Zoo made its first serious foray into the world of in situ conservation. A chance connection put me in touch with a group of Guatemalan conservationists who were building a rescue and rehabilitation center in the Peten, the northernmost rainforest region of Guatemala. After a visit there, we awarded them a five-thousand-dollar matching grant. With that initial grant, we tentatively placed our toes in the conservation waters, and we began to examine what our role could be in supporting on-the-ground conservation in other countries. What was the best and most effective approach? "These animals are the ambassadors for their cousins in the wild" was a frequently used refrain by zoos when speaking to their visitors, but to me those words were disingenuous if zoos were not actually committing their time and funds to conservation.

From 1990 to 2000 I actively became involved in formulating the philosophy, shaping the approach and building the Columbus Zoo's substantial commitment to field conservation. In 1996 I was named the first field conservation coordinator at Columbus, overseeing the numerous field and research projects we supported by then.

Recently I came across a paper I had written in 1998 for presentation at an Association of Zoos & Aquariums (AZA) conference. When I took a moment to read it, I was struck that although much had changed over the last twenty-eight years since Columbus began their initial commitment to conservation in 1990—many more zoos were providing start-up and long-term grants, there were more diverse zoo conservation models and partnerships with field people, and field researchers were now on staff at zoos and aquariums—the philosophy and formula that we had created that had built the Columbus Zoo's conservation reputation was still relevant today, especially for those zoos that were just beginning to initiate their own conservation commitment.

Columbus Zoo Formula, 1990

- Be humble; recognize we are not field researchers.
- Recognize that zoos have an infrastructure to offer.
- Empower staff and volunteers to be actively involved.
- Approach the project from a holistic standpoint.
- Contact a field researcher or a NGO representative; begin to develop a relationship with them.
- Offer some support, but don't make promises you may not be able to keep.
- Start with a small sum of money—small grants.
- Take your lead from the person in the field; don't presume to have the answers for what is best for that particular project.
- Build the relationship over time; remember your institution is proving itself as much as is the field researcher.

From this philosophy grew our commitment to a commonsense, holistic approach to supporting the people who were on the front lines of protecting wildlife. A significant part of this holistic approach was how we acknowledged the intrinsic value of the stories these people told. These stories were an essential part of their experiences expressing the challenges and triumphs they faced daily. But most especially, they shared stories of individual animals that affected them, that moved them and solidified their commitment to conserving those species. When telling their stories, the conservationist can elicit empathy from the listening audience and we zoos, because of our ability to collectively reach millions of visitors, became a powerful and useful conduit for their stories.

By 1998 I had been overseeing the conservation program at the Columbus Zoo for a number of years. Our financial and infrastructure commitment had grown exponentially since we started in 1990. We were known to our field colleagues as a responsive partner, easy to work with and innovative in our approach. We also were known for our diverse commitment to Great Ape conservation projects both in Africa and Southeast Asia.

Because of our far-reaching commitment, in 1997 I was asked to serve on the Amazon Conservation Team's (ACT) board, founded by ethnobotanist and author Dr. Mark Plotkin. My first board meeting took place in the Tucson desert in a compound with a collection of small rust-colored adobe cottages, each with a narrow front porch with overhead rafters covered in flowering vines. Outside my backdoor was a rose garden. Who would have thought that was possible in the desert? We were so far out in the country that at night I felt as if I could see the curvature of the sky, a dome of brilliant stars so numerous and bright that they illuminated the night all the way down to where earth meets sky.

We were a disparate group but all committed to conservation. One of the most notables was botanist Frits van Troon from Suriname. He was an imposing man, extremely tall with a kind face and gentle demeanor. His people, the Maroons of West African descent, were brought to South America centuries ago as slaves but found their way to freedom in the rainforests of Suriname. There were several shamans in traditional garb from Suriname and Colombia. My Columbus Zoo colleague Becky Rose was also there, and it was Becky's connection to Mark that brought his work to the attention of the Columbus Zoo. Finally, there was an elderly gentleman whom I met during our communal breakfast earlier in the day and who struck me as being a bit of a curmudgeon.

*   *   *

I am not normally a morning person. I like to ease into the day quietly and am a bit bitchy until I find my bearings, only after a good cup of coffee combined with time to myself. But the following morning, I rose early to walk the perimeter trails surrounding the compound to shake off my lethargy and travel fatigue. I saw the older gentleman in the distance, and I thought to myself, "How do I get around him without being rude?" I was looking for solitude not company. But as I approached him, we exchanged hellos and somehow fell into a companionable dialogue as we continued to walk. His name was Loren McIntyre, eighty years old, born in 1917, the year of my grandmother's birth, something that endeared him to me immediately. When he found out I had worked with gorillas, he lit up, his face animated as he told me of his capuchin monkey, ChiChi. He plied me with questions about ChiChi's behaviors he had experienced over the years, thinking I was an expert on all things primate. When we approached our bungalows to get ready for the meeting later that morning, I felt a little disappointed to be ending our conversation so soon but was also absolutely delighted to have found such an unexpected kindred spirit. I immediately felt comfortable with Loren, but I couldn't quite put my finger on why.

An hour later, our meeting started. Mark Plotkin welcomed us and then began the meeting by introducing Loren. Apparently Loren was a world-renowned National Geographic explorer, photographer, and writer. He was credited with discovering the headwaters of the Amazon River at seventeen thousand feet in the mountains of Peru. Loren stood up in his corduroy pants, lumberjack shirt, and white tennis shoes, tugging slightly on the bottom of his shirt, and it clicked: I finally knew what had made him feel so familiar. He reminded me of my father. He dressed like him and even carried himself a bit like Dad, his body lean and lanky with a long thin face so similar to my father's. I listened as he told us tales of his adventures, letting us know that the indigenous people always knew where the source of the Amazon River was, in a slightly chagrined voice. He updated us on his current projects. He spoke about why preserving these natural areas was so important.

Later when the board flew to Mexico to survey a community-based Seri Indian project the Columbus Zoo and ACT were supporting, Loren and I spent an evening in a restaurant talking endlessly of our lives—childhood, lost loves, our work. It became clear to me that no matter our age, no matter the lines on our faces, no matter our life experiences, both good and bad—at the core of each us is a hopeful child. And for like-minded souls like Loren and me, that kid was also fascinated by nature, by the outdoors, by the wonder of animals. In particular our mutual fascination and love of primates (he for ChiChi and me for Bongo) brought us together and solidified our friendship.

Several months later, I was in Washington, DC, for an evening event at the National Geographic headquarters. The following day I took the metro to Loren's house in Arlington, Virginia, and spent a wonderful afternoon and evening there with him and his wife, Sue. I was once again struck by his similarity to my father. While Sue and I sat on the couch talking, Loren took a seat on the floor, sitting with his back to the wall, legs bent at the knee. My childhood memories of my father include a vision of Dad sitting cross-legged on the floor reading a paperback book. I don't remember ever seeing him sitting on our couch or a chair unless it was at the dinner table. It was such an unusual thing for an adult to do, but obviously normal for Loren.

Sometime in late 1993 or early '94 I approached my boss, now assistant director Don Winstel, and we began discussing the possibility of doing a project, possibly hosting another Gorilla Workshop. Our discussion eventually morphed into the idea of hosting a conservation conference, one focusing on the potentially powerful role zoos could play by supporting field researchers and their projects.

I was proud that we were a zoo that was responsive to the needs of field researchers and conservationists. We did not over-committee to death our decisions. We did not make creating a philosophy and an action plan the main focus, but rather we focused on the actual implementation of support for field research. We were careful not to water down our intent, making it so generic that it lost its effectiveness, but rather we were willing to jump in and take chances, oftentimes with new or unknown projects.

In the mid-1990s many zoos still justified their existence through captive breeding programs alone. I am not saying captive breeding is not valid and a much-needed and recognized commitment on the part of zoos, but in our minds it was just a piece of the bigger puzzle. Captive breeding also struck me as being a shield—a fallback to hide behind when people called zoos on their lack of commitment to field conservation. It was a reflexive, defensive response, similar to blaming the gorillas for their supposed lack of social skills back in the '60s, '70s, and '80s when in actuality it was we humans who didn't understand what we were observing.

We were in our fourth year of supporting fieldwork and were constantly fine-tuning and adapting our approach, but what held true throughout was that actions speak louder than words—that we could not in good faith, or with any modicum of credibility, speak of conservation unless we (zoos) were proactive in our support. So we thought that perhaps by hosting a conference that forced the issue—addressing the role of zoos and how they were way behind in terms of their commitment to saving species in the wild—might be a positive start. Our intention was to gather field people and zoo personnel together to illustrate the many ways zoos could offer financial, infrastructure, and personnel support. And we wanted to acknowledge that field people have an expertise that we did not necessarily possess, but that zoos had an important and vital role to play by forming partnerships with and supporting field conservation.

In 1995 we hosted the inaugural Zoos: Committing to Conservation conference. When planning the conference, we had two very important underlying foci. We did not want scientific papers per se—we felt there were plenty of other conferences that afforded their telling. We wanted stories—stories that moved the field researchers, stories that would then move us (the zoo world) to action. We made it clear that we did not want speakers droning on about their data accompanied by graphs. Rather, we asked for speakers to tell us about something ordinary that had an extraordinary effect on them—something we could share with our public. We wanted stories of community-based conservation. We wanted to be inspired by people who just got on with what needed to be done, be they Rwandan, Congolese, Guatemalan, or Peruvian.

And just as importantly, we recognized that zoos and aquariums were in a unique position, distinct from other conservation organizations like the World Wildlife Fund, The Nature Conservancy, or Conservation International—we actually were a destination, a place people visited and, as such, we had an incredible opportunity through our signage, education programs, and websites to tell our field partners stories. We were in a position to enlist the help and support of our visitors.

Our keynote speakers were Dr. Russ Mittermeier of Conservation International; Dr. Mark Plotkin of Amazon Conservation Team; and the director of the National Zoo, Dr. Michael Robinson. Our fourth speaker was National Geographic photographer Michael "Nick" Nichols.

Worthington Industries was the corporate sponsor for our conference. My father, Joe Armstrong—a huge fan of Tim Cahill's writing and Nick Nichols's photography—worked for Worthington Industries as a factory worker. We paid a larger speaker's fee to Nick with the stipulation that he also had to give a lecture at our sponsor's company headquarters to their employees. Nick can get a lot of money for his lectures and our honorarium was still well below Nick's going rate, but because Columbus Zoo was such a big supporter of great ape field projects and conservation, I believe Nick willingly came to share his experiences with us at the most nominal of fees. Nick had spent the last thirteen-plus years documenting the plight of gorillas and chimpanzees in some of the most moving cover stories *National Geographic* had ever published. He was on a mission to tell their stories.

As Nick was walking across the conference hotel lobby, leaving to go give his talk at Worthington Industries, I called out, "Hey, Nick, say hello to my dad, Joe, will ya?" I didn't think anything more about it, until I stopped by my parents' home that evening. I asked Dad if he had seen Nick's lecture. My father, who has the gift of gab, who is a consummate BSer and is always ready with a witty quip, became strangely quiet, no smart-ass comments, nothing. Then he said, in a voice filled with something between wonder, pride, and disbelief, that when Nick took his place at the front of the packed corporate auditorium to begin his talk, Nick made this announcement: "Is Joe Armstrong here? I'm not starting until Joe is here."

I cannot even begin to explain what that meant to my father, to be singled out in a room full of his coworkers and I cannot fully express what Nick's gesture meant to me. My dad who was currently undergoing treatment for lung cancer, who taught me to love books and appreciate storytellers, who instilled in me an interest in photography because of his many black-and-white photos. My dad, who simply reveled in reading about Nick's many adventures with Cahill. I would be forever grateful to Nick for his kindness to my father. It was our mutual admiration, love, and respect for gorillas that brought Nick into my world and why I think he so generously and willingly gave his time to the Columbus Zoo.

In July 2000 the Columbus Zoo received a stamp of approval from the Association of Zoos & Aquariums (AZA) when we were reaccredited. This process involves a site visit of several days by a panel of zoo representatives from other institutions. They tour the zoo, examine exhibits for flaws and problems, interview staff, and look at animals' health records, signage, and annual reports. Reaccreditation happens every few years and is a big deal; suffice it to say, not getting accredited is a big black mark against a zoo. I recently found the four-page letter the AZA site team sent to us. The very last paragraph speaks volumes to the Columbus Zoo's commitment to field conservation:

> Conservation is the single most powerful message throughout the zoo. The Conservation and Collection Management Committee (CCMC) oversees the traditional planning process as well as oversight and approval for a plethora of research and field conservation initiatives. From somewhat humble beginnings in 1991 with a budget of $25,000 for conservation and research, the CCMC has grown to its current allocation of $280,000 (plus an additional $50,000 per year for the next five years from an anonymous donor). The level of funding has enabled the Columbus Zoo to implement and oversee this past year over 130 field conservation and research projects in Asia,

Australia, Central and South America, Africa, Mexico and North America.

I was extremely proud of what we had accomplished at Columbus, but an opportunity came my way when I was offered a job at a small Florida zoo, so six weeks after our AZA accreditation I left Columbus to embark on a new challenge, building a conservation program with limited funding and limited board support. But my new director, Margo McKnight, was a devoted conservationist as well, and we were both committed to creating a model of a small zoo as a conservation leader in the zoo world. And I brought with me my passion for great ape conservation, and as luck would have it my new director supported that passion.

Five years later, I found myself sitting in the hotel lobby of the Mount Kenya Safari Club in northern Kenya for a meeting of the Pan African Sanctuary Alliance (PASA)—a gathering of the managers of PASA primate sanctuaries established in as many as twelve African nations as well as the organizations and individuals who supported their work. It was a group of conservationists brought together by our mutual commitment to primates.

Mount Kenya Safari Club had seen better days; its former elegance a little rough around the edges, but that only seemed to make its charm shine through—its threadbare nature spoke of past residents, bygone events and experiences, and a past more glamorous. Previously owned by William Holden, the actor- and hunter-turned-conservationist, it was then owned by a wealthy Middle Eastern businessperson. Rumor had it that a new owner may be on the horizon, meaning possible upgrades and renovations. Rumor also had it that one of the current owner's relatives, might be making an appearance at the hotel during our stay.

After an early morning horseback ride, I showered and came down to the hotel lobby. It was a lovely ride, uphill through woodlands, then opening to savannas where I saw a variety of hoofstock, including a small herd of zebra. I enjoyed it so much that I already booked a second ride for the afternoon.

In the lobby, I sat down with a European colleague, and we became thick as thieves after discovering a mutual love of birding, although he is much more knowledgeable and devoted to the cause than I. I go birding more as an excuse to go for a walk, with little expectation of seeing anything, so I am inevitably thrilled whenever I catch sight of the winged gems that populate our world. We are planning an adventure for the following morning—to get up early before breakfast to have a wander about the place.

We were chatting away like like-minded spirits, planning our great birding expedition, blissfully unaware that a tempest was coming our way. An entourage arrived, burly men with stiff military bearing and gruff faces. They blew through the doors and swept through, looking left and right, taking our measure and obviously dismissing us as harmless or useless or both. Then a good-looking young man followed. He had the manner of the extremely wealthy—a princeling type—his demeanor spoke volumes of a sense of entitlement. He didn't acknowledge us either. We were nothing to him. "Ahhh," we said to ourselves, "the relative." My colleague and I did what any good student of primatology would do when encountering a new troop of gorillas or an unfamiliar group of chimpanzees or an interesting group of Homo sapiens: we settled in for the show.

Next came the girlfriend, wearing skintight pants and a blouse in a desert storm camouflage motif. On her feet was a set of soaring stiletto heels in camo as well. A Hermes bag or some sort of high-end purse hung from the crook of her arm. Her hair was long, sleek and black, so shiny it glistened. She was tall and extremely slender, willowy, gorgeous in an overly groomed way. My first thought was "I wonder what she looks like without all that makeup." My next thought: "How does she walk in those shoes?" She, too, didn't even deign to glance our way.

The next morning we enjoyed our leisurely stroll about the grounds in search of birds when we came upon a compound with high walls. We heard the distinct pant-hoot of a chimpanzee and then a none-too-happy European voice telling us to move away, that we were disturbing the animals. We looked at each other slightly startled as the disembodied voice hung in the air. This must be Karl Ammann, the Swiss wildlife photographer. In the early 1990s it was Karl who blew the cover off the bushmeat trade in great apes, as his photographs shook the conservation world with their brutality and honest depiction of the fate of these animals. His photos showed how the opening up of pristine forest areas via logging roads and the subsequent influx of workers looking for jobs and hunting of wildlife had moved from being locally sustainable to an untenable industrialized proportion. Eating gorillas and chimpanzees, once taboo in many cultures, had no cultural weight with strangers coming to these newly opened natural areas, and great apes were paying the price for it.

We met in Kenya because of the primate orphans left behind from this latest threat. The PASA sanctuary managers dealt with the aftermath of the killing of mother gorillas and chimps on a daily basis. The orphaned sons and daughters arrived at their facilities bruised and battered, both in body and spirit, traumatized by the cruelty of humans toward other living creatures. Their dedicated staff worked daily to build a lasting trust with these infants and juveniles—to bring them back from the abyss.

I became involved with PASA while at Columbus Zoo and carried that connection and commitment with me to my new employer in Florida. While at Columbus, I was blessed to have worked under the tutelage of Jack and then his successor, Jerry Borin, when we built the Columbus Zoo's conservation commitment. They believed in and empowered their staff to create, enact, and formulate philosophies that guided a vision of what good conservation was and could be. Columbus allowed and encouraged their staff to organize fund-raisers in support of fieldwork. They believed in the vision of their staff and volunteers. Without that environment, the reputation of Columbus's conservation commitment could not have grown, thrived, and become a model for other zoos by the end of the 1990s.

Nick Nichols sent an email to me on June 15, 2000, when he heard I would be leaving Columbus. He wrote, "Columbus Zoo has always been a leader in giving support to conservation projects in Central Africa. I have seen hard working field people stay alive with funds that they provide, always carefully and thoughtfully directed, the Columbus Zoo grants make a difference. The gorilla researchers at Mbeli Bai and the chimpanzee pilot study at Ndoki National Park (Republic of Congo) just might not exist without the support of Columbus, both are very important to science and conservation. Columbus Zoo is doing what all zoological parks should be doing—making a difference before it is too late."

My years of working in the field of conservation has shown me again and again that conservation can be complicated, it can be competitive, and it can be political, but at its core, conservation is actually quite simple—it is all about relationships. Conservation is about keeping your word, about delivering on your promises, about listening, responding, and working in partnership with your colleagues in the field.

Although it can be steeped in idealism, conservation is also very pragmatic. At its best, conservation approaches threats to wildlife and ecosystems from a holistic diverse standpoint, one that is based on community. Conservation is about local and indigenous communities, their voices, needs, and goals. Conservation is about more than animals; it is also about people having easy access to clean water, building medical clinics and schools, paying teacher salaries, and providing money for children's annual school fees. It's about building libraries and community centers. Conservation is about providing something as simple as sanitary napkins for African girls so they can continue to attend school during their monthly periods. And to those who think wildlife conservation is only about sentimentality, it is not. Bottom line is it all comes down to direct economic benefits—for those people who live on the front lines in small communities bordering national parks and protected areas; it is about providing a better life for their children.

49

TEENAGERS

Juvenile males are grubby and slightly disheveled. Everything about them seems a bit too big or ill fitting. The dirt of everyday life seems to find its way deep down through their hair coat, clinging to their skin so they always look a little gray, like they are in need of a thorough bath or hosing off. They are trouble with a capital *T* at the age of five, six, and seven.

Juvenile males seemingly suss out situations, calculate the trouble they can cause, and then have at it. They are masters of ornery, naughty for the sheer delight of it. And once again, because adults are so tolerant, only the most egregious antics will be dealt with.

I did not witness this event myself, but it was relayed to me by a fellow keeper. Apparently twenty-two-month-old Jumoke, newly introduced to Sylvia, was on the metal catwalk that is suspended high—really high, only about six feet below the mesh roof of the outdoor habitat. It's a handy walkway that meanders through the exhibit in a curvy, snakelike fashion. Gorillas can perch on it but that is rare. Most often it is used by females trying to get away from a bothersome and persistent male.

Because of its height, it is not really a place you want to see a lone infant. Sylvia was a good surrogate mom, but she had her limits—she wasn't overly doting, and seemed a bit absent-minded about the whole mothering thing at times. Jumoke was cruising along the catwalk when

FIGURE 49.1. Youngsters on catwalk in the habitat

not only did the keeper notice where she was, but so did the boys Mac and JJ as well—not good. It was almost as if you could see them rubbing their hands together in anticipation. Jumoke was oblivious to the two boys who were now hightailing it up the mesh, rapidly zeroing in on her. She was now hanging from the catwalk rung, playfully swinging her legs back and forth—apparently uncaring of the great drop below to the grass—when the boys arrived and began to methodically peel each of her fingers off the catwalk one at a time. She survived the fall unscathed, but it gave us quite a scare. Years later in 2007 I was in Rwanda to attend a primate conservation meeting. While there, I went to see gorillas in the wild—and I saw similar juvenile behavior.

It's not a long trek up to the gorillas. We walk through a small village, then terraced farmlands, and over a simple stonewall, the demarcation between private and public land, and finally cross over into the Parc National des Volcans in northwestern Rwanda, home to mountain gorillas.

We continue until we are enveloped by heavier vegetation. The morning air is cold. The undergrowth so thick and so springy underneath that a foot can easily slide down into the packed plants, which can reach all the way to your thighs; extricating yourself can be tricky. I had been to see mountain gorillas once, ten years earlier in Uganda. Now on this day in Rwanda, I am fortunate to once again trek to see gorillas.

We smell, then hear them before actually catching our first glimpse—their unmistakable pungent odor saturating the air before hearing the slow rumblings of a gorilla troop "talking" to one another. In a 1989 article in the *St. Louis Post-Dispatch,* explorer and conservationist J. Michael Fay described following lowland gorillas in the Congo Basin: "We knew we were close. The smell was so thick you could almost physically touch it."

No matter how many times I have been around gorillas, there is always that feeling of delight, coupled with a dose of amazement when in their presence. I look at the other people we are hiking with, all friends and colleagues, and we all share the same loopy smile—a "pinch me" sort of look on our faces.

When we finally see them, the gorillas are relaxed—eating, napping, and playing. The females nurse their babies while the adult male plays lord of manor, quietly watching over all. They are not even remotely curious about us. We keep our required distance due to the risk of spreading any type of infection we may unknowingly carry.

Our allotted time is up after an hour and we move into the bamboo forest. A well-defined path creates a tunnel-like corridor through this towering jungle, the sky overhead completely obliterated. The bamboo is so dense and thickly interwoven along the borders of the trail that there is no plausible exit off the path. I realize this when a juvenile gorilla makes an appearance—he has that cocksure strut, the "Hmmm, let's see what I can do to rile things up" attitude that all teenagers, humans and gorillas, can employ when of a mind. I see what's coming before he even begins. He's got his swagger on, walking stiff-legged, his insolence evident in his body language. I try to back into the tall bamboo quietly, but it is impossible to avoid him.

He does a half-hitch up and then down with his right leg—gathering forward momentum while beating his chest, "pok-pok-pok," before giving me a half-hearted thump on my leg as he runs by.

He is displaying, practicing at being a future leader. He's testing us, but with the arrogance of a youngster who is confident in his place in the world. He has the backing of his father, the silverback—and his mother and any other adult within the troop—should we do anything untoward or threatening toward him. He can bring the wrath of the entire troop down on our heads with just one cough-grunt from him. He knows it, and I know it.

The key to situations like this is to not react, to avert one's eyes, to relax your body and face. Although I have worked with gorillas for years, there is something slightly daunting when faced with a gorilla in the wild. Much to my relief, he carries on down the path, a jaunty attitude about him, just pausing for a moment to give me one last satisfied, slightly condescending look over his shoulder.

50

VOICES PAST
AND PRESENT

While mining ideas for this book in 2017, I pored through folders containing stories I had written over the years. I came upon an essay I had written in August 2015 about Fossey, which I promptly sent on to Ann Rademacher at the Little Rock Zoo where Fos had lived since 1993.

August 11, 2015

Everything about you was groundbreaking—your name, your birth, your childhood. You inherited your mother Bridgette's longish face with her sideburns, the unique shape of her nose, her signature red head. From your father Bongo, his distinctive rounded muzzle, deep-set grooved lines below rust colored eyes, his impressive physique and that thoughtful intelligent gaze. Well before your birth, we decided to name you Fossey. You became Fosman, Fos, Fossey to us.

You were born on a humid summer night, our first mother and father-reared gorilla at the Columbus Zoo.

Your father waited patiently nearby, quietly vocalizing to your mother Bridgette as she built her nest and gave birth to you.

The look on your dad's face the first time he touched you.

Your mom playing some sort of gorilla peek-a-boo with you as she covered you in hay, then in one big swoop lifted it all off, revealing your slightly puzzled face as if you didn't quite understand the rules of the game.

Your dad being goofy, doing double-rolls across the floor to land directly in front of you accompanied by a "mmm-wahh" vocalization, inviting you to play chase.

Your mom coming over and grabbing you in her arms if she thought the play session was getting too rough, your dad looking slightly shame-faced.

You followed your father everywhere. Everywhere.

The distinct sound of your laughter mixed with your dad's deep-throated chuckle.

The look of confusion and bewilderment on your face when your father passed away.

Watching you navigate yourself into a new troop—trying to find your place.

And when you moved to the Little Rock Zoo, I was forever grateful that you ended up there with a staff of like-minded keepers who loved you as we did, who saw your gentleness, your sensitivity, and who understood the monumental losses that shaped you.

In 2007, I received word that your female was pregnant and then got the news of the birth of your son, Mosi. Watching YouTube videos of you playing with him, you were Bongo all over again, his utter delight in your existence transferred now from you to your own son. In 2012, came word of the birth of another infant, your daughter Adelina.

Then late last night, fateful words scrolled across my computer screen as I checked emails, "Fossey: sad news." I simply could not, would not open it for the longest time, knowing what I would find, desperate to keep Pandora's box closed. Hot tears gathered, and then gently plopped on these pages smearing the ink as I try to write of your life today. I cannot seem to stop the unsolicited tears prickling behind my eyes, while thinking of you, of Bongo, your mother—your happy family, your Greek Tragedy of a life.

I dearly loved your father—his life in captivity represented all the poor decisions zoos had made for decades. Because of his life and yours, we made seismic changes in our philosophy, our husbandry approach. Gorillas were given the space to be gorillas; raising their young, interacting with one another, working out their own complicated social hierarchies. His life mattered more than I can say. And you, Fosman, were everything to him. At the end of his long and difficult life you gave him all that he'd never had—a beloved son. A shared bond so strong between you both that I cannot fathom that somewhere in this great vast universe, you and he are not at this very moment walking side by side, mirror images reflecting back your shared gentleness, your dignity, your willingness to be silly. Just catching up, you know, after so long apart—the inseparable team you always were.

Ann responded to my essay with her own memories and feelings about Fossey:

Beth, I was deeply moved by your essay, and it brought up such a rush of feelings. I reflected on how Fossey continued on his groundbreaking path for us, and me. He was my first youngster, and the father of the first gorilla born at our zoo. I was present at that birth and Fossey had learned well from Bongo, being watchful and respectful.

Many times I (and I believe I speak for all the ape keepers here at Little Rock) felt so grateful that we had Fos to be the father of our first-born. We have a picture of Fos and Bongo in our building, and when people marveled at what a playful, doting dad Fossey was, we always paid tribute to his upbringing and felt his skills could be attributed to lessons from Bongo.

Your descriptions of Bongo and Fossey together had me weeping and marveling at the same time. Truly, you described just what I saw watching Fossey and Mosi, including the double roll. I feel very connected to Bongo from the stories I'd heard and knowing that Fossey had so many of the things we loved about him modeled by his dad. Mosi's current keepers (at Brookfield Zoo) adore him and remark on what a steady, kind gorilla he is, so I know the legacy continues.

I feel a little bad for the guys that have to follow Fos. Our new silverback, Kivu, is a good guy and has posed no threat to

FIGURE 50.1. Fossey at the Little Rock Zoo.
Photo credit: Catherine Hopkins Tidwell, Little Rock Zoo.

Fossey's daughter Adelina. But, one of the other keepers said to me a few weeks ago, "He's no Fossey" and I knew just what she meant. I'm glad Adelina got some time with Fos as well. She was the first to approach Kivu and we could tell she really expected a game of chase or tickle. Thank you for sharing your essay with me. I remember as I was leaving with the boys in 1993 you said, "Take care of him. Remember he's an orphan." I never forgot.

In late summer 2010, I was asked to return to the Columbus Zoo Ape House as a consultant. Cassie was in the latter stages of her first pregnancy. It had been a number of years since the last gorilla birth and the current keeper staff had no hands-on experience dealing with newborns or, for that matter, a pregnant female.

Cassie's troop mates were nineteen-year-old Kebi and silverback Anakka. He had been pulled out and sat in the adjacent enclosure, keeping watch. His still sketchy behavior toward females continued to be somewhat unpredictable—even after all these years.

I spent several weeks prior to the birth charting Cassie's behavior, taking notes, typing them up at the end of the day for the staff, and meeting one-on-one with the keepers. I talked with them about what they should expect with a newborn.

As I sat in the old building on the very same cement bench that has been and continues to be the observation deck, I watched the keepers interact with the gorillas as they went about their daily tasks. Once again it was brought home to me that some keepers just have it, some intuitive sense, a grounded way of dealing with gorillas, and the gorillas, as always accurate in their ability to judge people's personalities, also know it. I heard it when Joe, an extremely tall keeper, entered from the kitchen area of the adjoining building. The gorillas start a chorus of welcoming vocalizations. They were happy to see him, and not because he was carrying a food bowl or treats, he was actually empty handed; they just liked the guy. They had a similar reaction to another keeper, a gal, because there was something gentle and unforced about her. You cannot teach that to a new keeper. You can teach a keeper to be efficient, to be timely, to watch and interpret certain behaviors, but there are particular keepers who instinctually know how to approach and work with gorillas. They see something within the troop and can translate a seemingly benign event correctly. They are able to put the pieces of the social jigsaw puzzle together with little or no effort. It is an innate gift.

I am once again blessed to be watching a newborn gorilla infant with its mother. First-time mom Cassie gave birth the day before while her troop quietly watched. The keeper staff has done an excellent job keeping to routine, and things are looking good for mom and baby.

Over the next days, I witness the barely contained excitement of "Auntie" Kebi who shares her elation with Cassie. I'm not sure what started it, but Kebi approaches Cassie, wraps her arms around Cassie's waist from behind, and they quick-walk around the perimeter of the enclosure, chatting enthusiastically, sometimes breaking into a bit of a run. They looked like a train, two compartments firmly attached, scooting along as one.

As I sit here watching the girls, it occurs to me that for thirty-plus years, on and off, I have been fortunate to have a glimpse into the lives of these fascinating creatures and their complicated social lives. Cassie shows all the signs of being a good mom. She is very matter-

FIGURE 50.2. Beth and Bongo. Photo courtesy of Beth Armstrong.

of-fact about this whole business of a baby clinging to her belly. She is her mother's daughter, after all. Sixteen years prior in this very same enclosure, Cassie herself was born to Pongi—symmetry at work.

As I watch Cassie and her infant daughter over the next six weeks, verifying nursings, documenting teething, and how mom and infant react to one another, I am struck by the physical similarities Cassie has to her father, Oscar, and to her grandfather, Bongo. When she hunkers down on her belly, casually glancing up at a passing keeper, she looks startling like her dad—it is his rust-colored eyes and mannerisms I see. And I had to wonder, where does she get the movements that so closely mimic her father who died before she was even born? Yet here he is looking up at me when she glimpses my way. Under her eyes are the furrowed lines and a hint of the rounded muzzle of her grandfather Bongo's face. Yes, she has her mother's sagittal crest, but in general she looks so much like her grandfather and father, it is as if they are here, still a part of both our lives.

These animals, from one generation to the next, bring with them the stories of their forefathers, of their daily lives, their individual personalities, their challenges and triumphs, their physical attributes.

I see it in the almond-shaped eyes of Cassie's newborn daughter, Nadami, eyes that perfectly mirror those of her father, Anakka.

These gorillas carry with them all that has gone before, from their torturous journeys from Africa to a captive setting at a US zoo. They carry the stories of their ancestors. As I sit in the midst of these sentient beings, I contemplate the timing of life's events and marvel at it all. For me, the journey with gorillas began years ago with a simple walk down the back aisle of the Ape House. It saved a young woman who was looking for direction and purpose.

AFTERWORD

At the 2013 Zoos and Aquariums: Committing to Conservation (ZACC) conference, a fellow zoo colleague stood up and said that they finally got it, that they realized conservation was all about relationships. My initial reaction was a sense of personal failure. Almost twenty years after the first ZACC conference, what was so essential to doing good conservation work was still a revelation to some, and that the core message had not been passed down to subsequent generations. And at that very moment, I heard a voice. Just as clear as day it said, "Mentor."

So I did. After the conference sessions ended that day, I approached six women—some worked at zoos, some in the field, some I knew, some I didn't, but each had a certain quality I was looking for, hard to quantify, but I knew it when I saw it. I gathered the few I did know, pointed out the others and asked them to have lunch together on the following day, while letting them know I couldn't join them as I had another meeting to attend.

The next day the women came to me, saying they had had an amazing lunch. I replied, "Here's the deal. I'm going to host a weekend at my house in Florida in the middle of winter, and you are all invited." A bed, all food and drinks, rides to and from the airport would be provided. The only thing they each needed to do was pay for their flights. So in January 2014, the first Conservation Mentoring Weekend took place. Since then, we have had three more. It is a weekend to share our stories, our accomplishments, the hurdles we face—sometimes on a daily basis. My hope was that this core group of young women would form an unspoken alliance, back one another and assist one another when things got tough, and that this next generation would finally shift the culture of zoos to become fully committed conservation organizations. That they would finish what we—the older generation—had started twenty-five years earlier.

In 2016 I was in a drugstore when I noticed several women's magazines on the stands. All had scantily clad women in somewhat provocative poses on the covers and it struck me, "This is it? This is what we offer up to the next generation of girls as models of success, of what success actually looks like? These are our heroines?"

I have been blessed to know countless women who are passionate about protecting wildlife, who are dedicated to working with local communities to fulfill their needs, building schools, libraries, and medical clinics halfway around the world all under the umbrella of conservation. I know women who get up every morning to conduct field research in difficult conditions, who work on issues that recognize and protect endangered wildlife, and who are doing everything within their power to make this world a better place. They are our heroines. They are the people who should be on the cover of magazines, inspiring us to do better, to be better.

Any single idea often organically grows into something unexpected and bigger, bigger than any one person and more far reaching. And that revelation in a CVS drugstore did too, seamlessly combining with what I had learned at the Florida mentoring weekends resulting in a conservation summit called Finding Your Voice: Inspirational Stories from Women Who Protect Wildlife and Wild Places.

By the spring, I had booked a meeting space for Finding Your Voice, secured commitments from most of my guest speakers, and then like so much in life, serendipity once again stepped in through a chance meeting with Dr. Anna Young, professor of the Zoo and Conservation Science program at Otterbein University in Westerville, Ohio. I had set up a meeting with Anna to discuss the possibility of me giving a lecture to her class about my experiences working with gorillas. I mentioned that I was hosting an event that her students might want to attend. Once I explained the premise, the where, what, and when, she said, without missing a beat, "Well, what if we host it here at Otterbein." And with that, we got up from her office, walked around Otterbein's beautiful liberal arts campus, looked at several potential meeting spaces, and that was it—a partnership was born. In October, 2016, we hosted the first Finding Your Voice with ten speakers.

When these little gems fall in your path—the revelation at the drugstore, the conservation weekends at my house, meeting with Anna—listen to them. I told the students attending this event, "That's the universe trying to get your attention. You may not know how these events are connected but take note. Set each little indicator aside, and if it's meant to be, another nugget will fall in your path that is somehow related and eventually all will reveal itself, a direction or revelation will come forth, and then you act on it."

Our speakers included several zoo representatives and several women who founded field projects in Madagascar, Uganda, and the Democratic Republic of Congo. We also had a graphic designer who was kind enough to create our logo and spoke about the power of art as a conservation tool.

When I asked our out out-of-town speakers to participate, not one of them balked for a moment when I told them I had no funding available to cover flights, but I would feed and house them while in Columbus. Eventually I was able to secure enough funds to cover their flights and allow for a small honorarium for each. But what was most telling is that they never hesitated, they simply said yes, generously willing to come on their own dime.

This two-day event was an opportunity to share our stories, our love of nature, of animals, and how each of us became involved in conservation. The focus on story made it possible to emphasize that there is no right or wrong way to get involved, that each of us had our own unique journey to travel, but no one path was more valid than another. What was important was to believe in yourself, follow your passion and intuition, work hard, volunteer, and recognize opportunities when they fell in your path—but just get there.

Several weeks before the summit, Dr. Jo Thompson, one of our speakers, called asking if she could bring some table decorations for her talk. "Absolutely!" was my reply. Jo explained how she was going

to create yellow, gold, and orange tissue paper "campfires" illuminated by battery-operated tealight candles that she would distribute
throughout the meeting room. She also planned to attach glittery silver stars to the walls. On the day of her talk, we lowered the lights and
the tealight campfires glowed while Jo spoke of the magic of sitting
around a campfire every night in the Democratic Republic of Congo
(DRC) with her Congolese colleagues—going over the day, sharing a
meal, sharing stories. Through her life's experiences and her evocative
storytelling, she transported all of us to the DRC, creating a sense of
camaraderie and community. She painted a picture so vivid we felt as
if we could smell the deep African night air with its ubiquitous wood
smoke and a chorus of insect and frog songs. It was pure magic.

Our hope was that the Finding Your Voice concept would go
regional, cropping up all over the country, thereby reaching more
students—both undergraduate and graduate. In 2018 Denver Zoo
hosted the second Finding Your Voice. Several of us from the first
Finding Your Voice gave lectures at Denver Zoo as well and, as more
and more audience members voiced their feelings, we were gratified
and somewhat incredulous at their responses. Many were moved to
tears, speaking of how much they needed this type of encouragement, that the meeting had affirmed they were on the right path and
they felt that they had finally found their "people."

In March of 2019 we hosted the third Finding Your Voice once
again at Otterbein University and have a confirmed host for the 2020
FYV in Omaha, Nebraska. I believe so strongly in the value of these
smaller, more intimate meetings—these gatherings of like-minded
folks are surely reminiscent of the kitchen table model. Our goal is
to provide a place to share our individual and communal experiences
with the next generation of conservationists, to hear their thoughts,
their challenges, and to then offer our advice and guidance. Then we
want them to go out and change the world.

In recent months, I attended a Gorilla Workshop where keepers,
curators, field researchers, and authors from around the globe gathered to share their experiences. This was the tenth workshop since we
had created it at the Columbus Zoo in 1990. And I was struck by a
change in tone perhaps influenced and seen through the lens of my
now much-older eyes. My generation of keepers were a product of
our time, informed by an entire generation of gorillas that had been

FIGURE A.1. 2016 Finding Your Voice speakers with the attendees.
Photo credit: Emma Parker Photography.

captured from the wild or had been pulled for hand rearing; therefore, we were fueled by the need for big transformative changes. We
felt passionately that we owed them recompense. We were fortunate
to be in the most unique, challenging, and exciting times—not informed by career advancement or by self-promotion but rather by a
sense of commitment to something bigger than us. And just as childhood acts as an influencing touchstone, informing how we navigate
through our adult lives, our unparalleled experiences with those particular gorillas shaped and molded us. And so we were the lucky few,
plopped down in the right place at just the right moment with just
the right people.

At the 2018 ZACC conference in Jacksonville, Florida, I was talking to
giant armadillo researcher Arnaud Desbiez about books. We were marveling about and singing the praises of the novel *A Story Like the Wind*
by Laurens van der Post, when Arnaud suddenly pointed behind me
toward the stage where many of my fellow ZACC steering committee
members were standing. I turned, thinking, "Huh, I thought we had

decided as a committee that we weren't going to have any formal talks while at the Zoo Day event but would just let people relax and network." I shrugged it off and turned back to Arnaud so we could resume our conversation. As anyone who knows me well will attest, a conversation about a good book is just about my most favorite topic in the world. But after another minute, Arnaud again pointed behind me and said, "Beth, I think they're calling your name." I had spent most of the afternoon off-site touring the White Oak Conservation Center north of the zoo and had arrived rather late to the zoo dinner, so I thought, "Well, I guess they decided to make some announcement of some sort while I was gone." So I dashed up to the stage, where, in addition to some of the ZACC Steering Committee members, I noticed many of my colleagues who have participated in the conservation mentoring weekends as well as the Finding Your Voice summit. They stunned me by presenting me with a beautifully embossed book of inspirational quotes, poems, and testimonials from friends and colleagues—accompanied by photos I had taken over the years. Then they announced that they had established an education scholarship for Malagasy children in my name. That truly took my breath away.

This mentoring, this sharing of experiences and stories winds its way back to Bongo and his influence over and over again. I opened up the first Finding Your Voice summit with a photograph of Bongo. I said to all the young women and men in the room, their expectant and idealistic faces turned toward the screen, "Mentors come in all shapes and sizes and, in my case, species. Everything, absolutely everything loops back to Bongo." He was and is my constant, my North Star in all that I do.

ACKNOWLEDGMENTS

Thank you to all the people at The Ohio State University Press/Trillium for making this book happen—Tony Sanfilippo, Tara Cyphers, Laurie Avery, Debra Jul, and Samara Rafert. Thank you for your kind words, encouragement, and patience and most especially for believing in the gorillas and the stories they had to tell. Thanks to Kathy Wallace for your guidance and input. Thank you to Melody Negron at Westchester Publishing Services for the beautiful layout of the book.

Thank you to the Columbus Zoo and Aquarium—as an institution you were a leader on so many different levels. It was a joy and a privilege to work at such an innovative and nurturing zoo. I have been and will forever be proud to say I once worked at the Columbus Zoo.

Thank you to my fellow gorilla keepers, most especially to Charlene Jendry. Our in-depth discussions resulted in something bigger, something that changed the face of captive gorilla husbandry. To Adele Absi (Dodge) for your unique ideas, consistent reliability, and your never-ending work ethic; we were a great team together. And a special thanks to Dianna Frisch for hiring me and giving me my first glimpse of gorillas. We all worked together in a halcyon time, days of such unrestricted creativity that I think we were intoxicated by the endless possibilities.

To Nancy Staley, you were our fifth keeper. Your commitment and quiet presence allowed for the beautiful photographs and videos you took over those crucial years, documenting monumental changes. This book would have been a shell without your images adding to the stories. Thank you.

A special thanks to Susan White and Debbie Elder who later joined us and continued to push the program forward after the original four keepers had all departed, and to Liz Garland and Laura McMahon, our two seasonal keepers. All of these women got little of the recognition or glory, but they were instrumental and essential at a time of enormous changes within the Ape House.

To Don Winstel who hired me that fateful spring day. You and I have had numerous conversations over the years about the reasons why so much was accomplished then. I think it was one of those magical moments when the stars were aligned and every element was in place allowing for change to occur. What you did, Don, was listen to us. You supported our decisions, questioned us when perhaps we needed it, and most importantly, you trusted us—our judgment, our commitment to the gorillas. Maybe it was the push and pull, the sometimes butting of heads among ourselves, the keepers, but also with you at times. All of it resulted in making the necessary changes that had been long overdue.

To Jack Hanna, who created such a rich and creative environment that his staff felt as if they could do anything, that they could accomplish anything. You allowed us to be the drivers in improving the lives of the animals within our care and empowered us to make a difference in limitless ways. It resulted in animals living rich, complicated social lives. Your faith in your staff was a gift beyond all measure. We were spoiled.

To Jerry Borin, our zoo director who took over after Jack became director emeritus. Your kind and quiet approach, your lack of ego, and continued support for what started under Jack's watch was seamless. You made it easy for our passion and commitment to our programs to continue and flourish without a hitch. We were spoiled twice.

To Dusty Lombardi, my fellow keeper who spent countless hours watching snow monkeys with me, sharing many conversations about primate behaviors. Who taught me to multitask, to be both organized and adaptable—and most importantly who supported my decision to study at Apenheul in the Netherlands.

To Bill Cupps, head keeper of cheetahs and bears, you instilled in me the belief that the animals themselves would give us the answers to solving behavioral problems. And thank you for never once begrudging my love of working with gorillas even if it meant taking me away from your section.

To Kate Oliphint, a simple thank you seems woefully insufficient.

To Doug Cress, who once told me I should write a book about Bongo. You may not recall that but I never forgot it and that comment and encouragement was a seed that grew.

To my father, Joe Armstrong, who asked how Bongo and Fossey were doing after the death of Bridgette, who listened and reveled in the many gorilla stories I brought to him. My father taught me to love books, to look at life from another person's perspective, and to love animals.

To my grandmother (Mom-mom), Pauline Shuler Authenrieth, who always supported my choices in life no matter how seemingly odd or off-kilter. You loved hearing about the gorillas and fondly asked after them as if they were old neighbors. Your love was pure and always unequivocal.

To my mother who gave me one of the greatest gifts of all— books—taking me to the library every Thursday evening and allowing me the freedom to read anything and everything. My mom, who sat up late and fed a baby bird.

To my brother, Fred, and his wife Mary Jo. Fred, when I once sent you a story I had written about our childhood, Mary Jo immediately called back to tell me you both were crying. Thank you both for reading it, giving me the confidence to write more stories.

To my monthly International Women's Writing Guild (IWWG) group: you have encouraged, supported, and set me straight when I needed it. I am not much of a joiner, but I have found these monthly meetings to be a necessity in my life. Thank you, Jeanne, Doris, Jules, Bonnie, Val, Nice, Alice, Linda, and Clay for your always gentle but firm and constructive suggestions.

To Nick for proofreading over and over again. Thank you for your patience and support. And to Nick's sister, my sister-in-law, Caroline Mason (Bebe), for lending your discerning eye to proofread the original essays years ago. Those essays were the seeds that took root resulting in this book. I cannot thank you both enough.

Thank you to Jan and Tom Parkes for capturing such a poignant image and allowing us to use it for the front cover. And to my talented niece, Emma Parker, for taking the photo of the original Finding Your Voice crew. That photo means the world to all of us as we recently lost one of our members.

And to the gorillas themselves but to Bongo in particular—thank you. You taught me what poise and dignity look like, what love looks like in another species—not so very different from our own. You taught me humility. You taught me what purpose is. You graced me with your presence. Your history, your heartache, your joys all mattered. Your life mattered.

INDIVIDUAL PROFILES

Adult Males (Silverbacks)

Anakka (1985–2016): Male, born at the Philadelphia Zoo. Anakka came to the Columbus Zoo as a blackback at the age of eight with his nine-year-old brother Chaka in 1993. Both very disruptive because of their young age, they were far too immature to lead a troop. Anakka sired several offspring while at Columbus.

Baron Macombo (1946–1984): Male, wild-caught. Mac, as he was known, was brought to Columbus in 1950 along with Millie. Father to Colo. He was a solitary male when I met him but had been Millie's mate prior to her death in 1976. He was very calm, rarely vocalized, and had poor vision.

Bongo (1956 or 1957–1990): Male, wild-caught, brought to Columbus Zoo in October 1958, as a companion to Colo. Bongo could be both regal and goofy. He was highly respected by all who knew him, gorillas and people alike. Bongo was extremely handsome, very gentle and had a great sense of humor. Father to Emmy, Oscar, Toni, and Fossey. He raised his son Fossey after the death of his mate, Bridgette.

Mumbah (1965–2012): Male, wild-caught, brought to Columbus in 1984 from Howletts Wild Animal Park in the UK. Nicknamed Mums or Mr. Mumbah. He was a surrogate father, allowing us to create a troop of adoptive parents. We would not have been able to enact the Gorilla Surrogacy Program without him. He was quiet, kind of removed, and extremely tolerant to changes within his troop. Mumbah was a calm presence, serving as a father figure to numerous infants and juveniles.

Oscar (1969–1993): Male, born at Columbus, son of Colo and Bongo, brother to Emmy and Toni. He led his own troop in the north building and fathered numerous offspring, but it wasn't until 1987 that he was given the opportunity to raise his own offspring, Colbridge. He was a wonderful father.

Sunshine (1974–2008): Male, captive born, from San Francisco Zoo. He was extremely tall. Sunshine, also called Sunny or Shine, was a little rough on his females when he first arrived at Columbus, but he eventually matured and developed a more measured approach to wooing his girls. He sired many infants while at Columbus, and all of his offspring had distinctive pink pigmentation (either marbling or entire pink digits) on their hands and feet, or both.

Adult Females

Bathsheba (1957–1998): Female, wild-caught, came from Cheyenne Mountain Zoo. Bathsheba, also known as Bath or Bathie, was the surrogate mother to Nkosi. She had arthritic hands and feet so they were always curled in.

Bridgette: (1961–1987): Female, wild-caught, brought from Omaha's Henry Doorly Zoo in the early 1980s. Mother to the twins, Mac and Mosuba, who were born in 1983 and to another male Motuba, born in 1985, all sired by Oscar. Then Bridgette became Bongo's mate. She and Bongo became parents to Fossey in August 1986 and they raised him together. Bridgette died when Fossey was fourteen months old.

Colo (1956–2017): Female, first gorilla born in captivity, born at the Columbus Zoo. Colo was a mate to Bongo. Colo's parents were Millie and Mac, but she was raised by humans in the nursery. Colo, nicknamed Granny later in life, loved wearing things on her head and drinking out of the hose—and she was a spitter. The Columbus Zoo's first surrogate mother, she adopted her grandson JJ when he was fourteen months old. Colo was also the biological mother to Emmy, Oscar, and Toni.

Cora (1979–2003): Female, born at the Columbus Zoo to Toni and Oscar. She was the infant I saw on a television program that piqued my interest in working with gorillas. In the early days of forming the surrogate troop, she was an instrumental member. Nicknamed Cora-belle, she was the first adult gorilla that the twins went in with.

Lulu (1964–2011): Female, wild-caught, brought to Columbus in 1984 from the Central Park Zoo. Lulu raised her daughter Kebi Moyo at the Columbus Zoo. She had raised another daughter Patty Cake while at the Central Park Zoo. Sometimes called Lu or Boo, Lulu was also a surrogate mother to her granddaughter, Kambera, and to a male infant, Umande.

Millie "Christina"(1949–1976): Female, wild-caught, brought to Columbus in 1950 along with Baron Macombo. She was named Christina in a "Name the Gorilla" contest but eventually was called Millie after the zoo director's wife, Mildred. Millie was mother to Colo (sire was Baron Macombo, known as Mac).

Muke (1965–2009): Female, wild-caught, owned by Saint Louis Zoo. Extremely large for a female, weighing in at approximately 275 pounds. Brought to Columbus in 1982 but left in 1984 to go to several other zoos before finally settling at Utah's Hogle Zoo in 1996.

Pongi (1963–2014): Female, wild-caught, brought to Columbus in 1985 from the Birmingham Zoo. Pongi (pon-JEE) was featured in a segment of *The Urban Gorilla* documentary and was a high-ranking matriarch who raised her own biological young and acted as a surrogate mother to other infants.

Sylvia (1963–2004): Female, wild-caught, arrived in Columbus in late 1986 from the National Zoo but owned by the Baltimore Zoo.

Sylvia had a palsy-like shake to her hand and face, especially when stressed. She was surrogate mother to Jumoke, Nia, and Little Joe, and an auntie figure to many other youngsters.

Toni (1971–): Female, born at Columbus. Toni was the last of Colo and Bongo's kids; her siblings were Emmy and Oscar. Toni (aka Toni-Baloney) was Sunshine's favorite female. She had numerous offspring but did not raise them herself; her kids were adopted by the surrogate group. Toni was shipped to the Detroit Zoo with Sunshine and Cora in 1996 but returned to Columbus in 2008 after the deaths of Cora and Sunshine. As of this writing, she is still at the Columbus Zoo and is currently a surrogate mother in Macombo's (the twin) troop.

Juveniles and Infants

Binti Jua (1988–): Female, born at Columbus to Lulu and Sunshine. Binti Jua means "daughter of Sunshine" in Swahili. She was pulled from her mother at almost two months of age due to the lack of weight gain. Eventually, she was sent to Brookfield Zoo where she became world famous in 1996 when a human toddler fell into her exhibit and she cradled him gently until zoo staff arrived.

Casode (1993–): Female, born at Columbus to Oscar and Pongi. Nicknamed Cassie, she was born into and raised by her mother Pongi in Mumbah's surrogate troop after the death of her father, Oscar. Cassie became a mother in 2010 (her sire was Anakka) and raised her daughter, Nadami.

Colbridge (1987–1994): Male, born at Columbus. He was routinely called by his nickname, Colbi. Raised by his mother Pongi and father Oscar, Colbi was the second infant at Columbus Zoo to be raised by his parents. Colbi was in the 1990 documentary *The Urban Gorilla*.

Fossey (1986–2015): Male, the first infant at Columbus reared by both parents, Bongo and Bridgette. Fossey, sometimes called Fosman or Fos, was an extremely gentle and kind gorilla who moved to the Little Rock Zoo in 1993 along with his close friend JJ. Fossey fathered and helped raise two infants while at Little Rock, a male named Mosi (now at Brookfield Zoo) and a female named Adelina (still at the Little Rock Zoo).

JJ (1987–2008): Male, born at Columbus to Sunshine and Toni. JJ (Jungle Jack—named after Jack Hanna as JJ was born on his birthday) was pulled for nursery rearing after his mother did not show an interest in raising him. At fourteen months, he was the first infant integrated into the surrogate troop. His grandmother, Colo, adopted him. JJ, sometimes called J or J-Bird, was mischievous—always looking for trouble—but in a friendly way. He and Fossey hit it off and were constant companions.

Jumoke (1989–2008): Female, born at Columbus to Toni and Sunshine. Exceptionally beautiful, Jumoke (aka Mok or Mokie) was our second nursery-reared youngster integrated into the surrogate troop. Sylvia adopted her.

Kebi Moyo (1991–): Female, born at Columbus to Lulu and Sunshine. Kebi was raised in Mumbah's surrogate troop where she was born. Very indulged by her mother, she became good friends with Cassie, Pongi's daughter. She gave birth to Kambera, via C-section. Lulu, Kebi's mother, ended up adopting and raising Kambera. Kebi currently resides at the Cleveland Metroparks Zoo.

Macombo II (1983–) and Mosuba (1983–): Male twins born at Columbus to Bridgette and Oscar. They were the first true youngsters integrated into the surrogate troop. Macombo II, or Mac (named for his grandfather Baron Macombo) is still at Columbus, and sired his first offspring in 2016. Mac has also been a surrogate dad to other infants. The twins were separated at age seven when Mosuba was taken to the Henry Doorly Zoo in Omaha, Nebraska. Mosuba's semen was used to sire the first test-tube western lowland gorilla baby, born at the Cincinnati Zoo in 1995. He is now at the North Carolina Zoo where he is a surrogate father to Nkosi's kids, Apollo and Bomassa.

Nia (1993–): Born at the Oklahoma City Zoo, brought to Columbus in 1994. Nia was a game-changer because she was the first infant sent to us from another zoo to be integrated into Mumbah's surrogate troop. Sylvia was her adoptive mother—the second infant Sylvia adopted. Nia still resides at the Columbus Zoo.

Nkosi (1991–2013): Male, son of Toni and Sunshine, nursery reared. Nkosi, which means "leader" in Swahili, was named after Dr. Nick Baird, ob-gyn to the gorillas. At eight months of age, Nkosi, also called Nikki or Nik, was the youngest infant integrated into Mumbah's troop at that time using a surrogate mother, Bathsheba. He died in 2013 at the North Carolina Zoo, leaving behind two offspring Apollo and Bomassa (see Mosuba bio).

GORILLA COMMUNICATIONS

Facial Expressions

- Lips pulled in, pursed: annoyed
- Sidelong glance: annoyed or aggressive. Lips pulled in, and a sharp glancing movement from the side, looking at the source of their displeasure. The glance may be accompanied by a stiff-arm/leg stance. It also can be accompanied by a chest beat or run-by display.
- Play face: happy, playful. This open-mouth grin is usually accompanied by laughter but not always.
- Infant frightened: lips puckered, forming an *o* with a whimper, like a hooting cry.
- Yawning: can indicate nervousness, tiredness, or potentially a warning.

Body Language

- Chest beating: can be aggression or just goofing around in play—it depends on the facial expression and other body language. Beating cupped hands against the chest in a rapid-fire motion creates a *pok-pok-pok* sound.
- Chest beating with stomping of feet: could be a play display, depending on facial expression and vocalization.

- Chest beating with foot stomping combined with a run-by: if accompanied by a stiff-legged stance, pulled in lips, a hoot series vocalization, and sidelong glances, this is a very annoyed gorilla making a point to another.
- Stiff-legged stance: arms/legs extremely stiff, body tense, usually accompanied by sidelong glances at the object of their annoyance. The stance can sometimes be used by a female who wants to breed with a male.
- Run-by: is just that, running by or near another troop member(s), usually accompanied by chest beating before or after the run-by display. Can be a dominant move—making a point. Sometimes used by a male trying to break up a possible altercation—inserting himself in the middle of an issue within the troop in order to deescalate it. Used by both males and females.
- Turning their back with a soft look over their shoulders: an invitation to play or chase.

Vocalizations

- Belch vocalization: described by Dian Fossey as *naoom* throat-clearing sound, it sounds like an intake of breath, *na,* followed seamlessly by a deep humming rumbly sound, *naa-hummm.* The vocalization also can sound like *mmm-wahh* depending on the gorilla. It indicates that life is good, a sense of contentment. It's also used to tell others "Here I am" or to ask "Where are you?" Both males and females use this vocalization. It is the most commonly used.
- Feeding vocalization: varies from gorilla to gorilla, but sounds like a rumbling culminating in a quick intake of breath several times. Males and females of all ages use it to display excitement.
- Ha/harummph: an invitation to play, interact, chase in play, or engage in tickling.
- Laughter: a deep chortling sound. Gorillas of all ages laugh. Infants can be made to laugh by tickling them in their groin, under the armpits, and under their necks. It is an infectious sound.

- Cough-grunt: a short burst of cough-like sounds used as a warning to back off of food, move, stop picking on your sibling, or cut it out. It sounds like *oo-oo-oo* (rhymes with *new*) and can either be soft when directed toward an infant or juvenile, or louder when directed at an adult. The cough-grunt should *never* be used by an infant or juvenile to an adult—it is akin to us as children using a disrespectful or challenging tone of voice when speaking to an adult.
- Alarm bark: described by Dian Fossey as a *wraagh* sound. It is short and sharp, extremely loud and piercing, and very distinct. The male uses this when the group may be in danger.
- Scream: females in particular use it when a fight is starting, escalating from cough-grunts or during an actual fight.
- Hoot series: usually used when there is a distance between group members and then they come closer together, or when winding up for a run-by display.
- Mourning call: a very mournful sound that starts low in pitch and ascends higher and higher, ending in what sounds like a question. The mourning call is used to call for lost members, those that have left the troop or died.
- Breeding solicitation: a reverberating summons, which sounds like a question.
- Infant whimper: lips puckered in an *o* with a *hoo-hoo-hoo* sound when frightened, insecure, unsure, or in pain. This *o* sound can turn into a raspy whining vocalization, a worried insecure sound like an elongated whimper. The whimper can culminate in a scream if physically hurt or extremely frightened.

TROOPS

Some troops were very fluid, with members changing groups frequently.

Bongo's Troop

Bongo	male	1956/57	wild born
Bridgette	female	1961	wild born
Fossey	male	1986	Columbus Zoo

Sunshine's Troop

Sunshine	male	1974	San Francisco Zoo
Toni	female	1971	Columbus Zoo
Lulu	female	1964	wild born
Cora	female	1979	Columbus Zoo
Molly	female	1975	Kansas City Zoo

Oscar's Troop

Oscar	male	1969	Columbus Zoo
Pongi	female	1963	wild born
Sylvia	female	1963	wild born
Colbridge	male	1987	Columbus Zoo
Muke	female	1965	wild born
Bridgette	female	1961	wild born

Mumbah's Troop—Surrogate Group

Mumbah	male	1965	wild born
Colo	female	1956	Columbus Zoo
Cora	female	1979	Columbus Zoo
Macombo (twin)	male	1983	Columbus Zoo
Mosuba (twin)	male	1983	Columbus Zoo
JJ	male	1987	Columbus Zoo
Bathsheba	female	1957	wild born
Sylvia	female	1963	wild born
Lulu	female	1964	wild born
Kebi Moyo	female	1991	Columbus Zoo
Pongi	female	1963	wild born
Colbridge	male	1987	Columbus Zoo
Casode	female	1993	Columbus Zoo
Jumoke	female	1989	Columbus Zoo
Nia	female	1993	Oklahoma City Zoo
Nkosi	male	1991	Columbus Zoo
Fossey	male	1986	Columbus Zoo

MOTHERS, FATHERS, AND INFANTS

Biological Mothers, Fathers, and Infants

Mother	Father	Infant	Birth Year
Colo	Bongo	Emmy	1968
Colo	Bongo	Oscar	1969
Colo	Bongo	Toni	1971
Bridgette	Bongo	Fossey	1986
Toni	Oscar	Cora	1979
Toni	Oscar	Kahn	1980 (died)
Toni	Oscar	Zura	1981
Pongi	Oscar	Mwelu	1986
Pongi	Oscar	Colbridge	1987
Pongi	Oscar	Casode	1993
Bridgette	Oscar	Macombo	1983
Bridgette	Oscar	Mosuba	1983
Bridgette	Oscar	Motuba	1985

(continued)

Mother	Father	Infant	Birth Year
Toni	Sunshine	JJ	1987
Toni	Sunshine	Norman	1988 (died)
Toni	Sunshine	Jumoke	1989
Toni	Sunshine	Nkosi	1991
Lulu	Sunshine	Lusi	1987
Lulu	Sunshine	Binti Jua	1988
Lulu	Sunshine	Kebi Moyo	1991
Molly	Sunshine	Unnamed	1992 (died)
Molly	Sunshine	Unnamed	1993 (died)
Jumoke	Anakka	Jontu	1997
Jumoke	Anakka	Muchana	2000
Casode	Anakka	Nadami	2010

RECOMMENDED READING

Recommended Reading
Gorillas in the Mist, Dian Fossey
Gorilla in Our Midst: The Story of the Columbus Zoo Gorillas,
    Jeff Lyttle
Jambo: A Gorilla's Story, Richard Johnstone-Scott
A Wolverine Is Eating My Leg, Tim Cahill
Eating Apes, Dale Peterson with photographs by Karl Ammann
A Wild Life: A Visual Biography of Photographer Michael Nichols,
    Melissa Harris
Demonic Males, Richard Wrangham
Bonobo Handshake, Vanessa Woods
In the Shadow of Man, Jane Goodall
Visions of Caliban, Jane Goodall
Chimpanzee Politics: Power and Sex Among Apes, Frans de Waal

Children's Books
Little Beauty, Anthony Browne
Gorilla, Anthony Browne
A Mom for Umande, Maria Faulconer
A Sick Day for Amos McGee, Philip C. Stead
Mountain Mists: A Story of the Virungas, Evelyn Lee
*Looking for Miza: The True Story of a Mountain Gorilla Family Who
    Rescued One of Their Own,* Juliana Hatkoff, Isabella Hatkoff,
    Craig Hatkoff, and Dr. Paula Kahumbu
Gorillas Up Close, Christena Nippert-Eng
Gorilla Doctors: Saving Endangered Great Apes, Pamela S. Turner
Eyewitness Books: Gorilla, Money & Ape, Ian Redmond
Eyewitness Books: Gorilla, Ian Redmond
The Heart of the Beast: Eight Great Gorilla Stories, Nancy Roe-Pimm

HELP PROTECT WILD GORILLAS

If you would like to contribute to protecting gorillas in the wild, these three projects are each in their own way making a huge difference. Please consider donating to any or all of them. And realize that what you may consider to be a small amount of money can actually go a long way with these projects on the ground. You can truly make a difference, and thank you for caring about gorillas!

Ape Action Africa
Where: Cameroon, West Africa

To provide sanctuary for individual endangered primates orphaned by the illegal bushmeat and pet trades; to actively rescue orphaned primates, providing rehabilitation and long-term care in a protected environment; and to work closely with the Cameroonian people to protect their natural heritage through education and social support.

For more information, go to http://www.apeactionafrica.org/index.

To donate, go to http://www.apeactionafrica.org/donate/donate.

Partners in Conservation, Columbus Zoo
Where: Rwanda and the Democratic Republic of Congo

By providing resources and funds to local conservation leaders, Partners In Conservation and the Columbus Zoo put money to work on the ground to advance conservation research, build capacity among local populations, strengthen environmental education, promote community involvement, and foster behavior change in Rwanda and the Democratic Republic of Congo.

For more information, go to https://globalimpact.columbuszoo.org/about/partners-in-conservation.

To donate, go to https://give.columbuszoo.org/partners-in-conservation.

Mbeli Bai Western Lowland Gorilla Study,
Wildlife Conservation Society
Where: Nouabale Ndoki National Park, Republic of Congo

The Mbeli Bai Study is the longest running study on the social organization, life history and demographics of Western Lowland Gorillas in the wild—initiated by researchers from the Wildlife Conservation Society in February 1995. The study has been monitoring the large mammals visiting this thirteen-hectare forest clearing (bai in the local language) from an observation platform at the site's edge for almost twenty-five years.

For more information, go to https://www.mbelibaistudy.org/gorilla.

To donate, go to https://www.mbelibaistudy.org/get-involved.